My Book

This book belongs to

Name: _____

Cover Design by :
Gowri Vemuri

First Edition :
June, 2020

Author :
Gowri Vemuri

Editor :
Ritvik Pothapragada

Questions: mathknots.help@gmail.com

This book is dedicated to:

My Mom, who is my best critic, guide and supporter.

To what I am today, and what I am going to become tomorrow,

is all because of your blessings, unconditional affection and support.

This book is dedicated to the

strongest women of my life ,

my dearest mom

and

to all those moms in this universe.

G.V.

www.math-knots.com

www.math-knots.com

Advanced Algebra 1 notes

Expressions	Numerical expression	Algebraic/Variable expression
Expressions will not include equal sign	12^2 $8 + 10$ $4^2 - 16 + 9$	$6x + 17$ $5a^2 - 2b$
Equation	**Numerical equation**	**Algebraic/variable equation**
Must include an equal sign; One side is equal to the other side	$9 + 8 = 17$ $5^2 - 11 + 2 = 16$	$6x + 5 = 35$ (x=__?__) $9x^2 = 225$ (x=__?__)

Solving Equations :

To solve an equation first add the like terms (if any) and then isolate the variable on to one side of the equation. To solve one - step equations, do the inverse operation to find the value of the variable.

Remember : Always perform the same operation on both sides of the equation to maintain the balance.
Inverse operation for addition is subtraction and vice versa.
Inverse operation for multiplication is division and vice versa.

Polynomial	Highest degree	1st Name by degree	2nd Name by number of terms
8	0	Constant	monomial
$9x + 8$	1	Linear	binomial
$-2x^2 + 7x - 11$	2	Quadratic	trinomial
$8x^3$	3	Cubic	monomial
$2x^4 + x^3$	4	Quartic	binomial
$3x^5 + 5x^3 + 12$	5	Quintic	trinomial
$2x^6 + x^5 - 3x^4 - 21$	6	6th degree	Polynomial with 4 terms
$5x^7 - x^5 + 4x^4 - 7x^2 + x - 3$	7	7th degree	Polynomial with 6 terms

A Polynomial in standard form is always written with terms in sequential order according to their highest degree of exponents (higher exponents to lower exponents).

Parts of the exponent:

7 is the base

4 is the exponent

This is read as "Seven to the fourth power"

Like Terms :

Two or more terms are said to be alike if they have the same variable and the same degree. Coefficients of like terms are not necessarily be same.

An expression is in its simplest form when

1.   All like terms are combined.
2.   All parentheses are opened and simplified.

Like Terms can combined by adding or subtracting their coefficients (pay attention to the positive and negative signs of the coefficient and apply rules of adding integers)

Combining like terms on the opposite side of the equal sign :

When the like terms are on opposite sides, we have to combine like terms by using the inverse operation and by undoing the equation.

Solving equations using the distributive property :

The number in front of the parentheses needs to be multiplied with every term within the parentheses. After the distribution and opening up the parentheses, combine like terms and solve.

Distributing with the negative sign :

Remember to apply the integer rules of positive and negative numbers while distributing.

$$+ \times + = +$$
$$- \times - = +$$
$$- \times + = -$$
$$+ \times - = -$$

www.math-knots.com

Example 1 : $2x + 3 = x + 7$

$$2x + 3 = x + 7$$
$$\underline{-x \ -3 \quad -x \ -3}$$
$$x + 0 = 0 + 4$$
$$x = 4$$

> Inverse operation for addition is subtraction

Example 2 : $7x + 5 = -3x + 25$

$$7x + 5 = -3x + 25$$
$$\underline{3x \ -5 \quad 3x \quad -5}$$
$$10x + 0 = 0 + 20$$
$$10x = 20$$
$$\frac{10x}{10} = \frac{20}{10}^{2}$$

> Inverse operation for addition is subtraction and vice versa

> Inverse operation for multiplication is division

$$\boxed{x = 2}$$

Example 3 : $\dfrac{2x}{5} + 5 = 15$

$$\frac{2x}{5} + 5 = 15$$
$$\underline{\qquad -5 \quad -5}$$
$$\frac{2x}{5} + 0 = 10$$

$$\frac{2x}{5} = 10$$

$$\cancel{5} \cdot \frac{2x}{\cancel{5}} = 5 \cdot 10$$

$$\frac{\cancel{2}x}{\cancel{2}} = \frac{\cancel{50}^{25}}{\cancel{2}}$$

$$\boxed{x = 25}$$

> Inverse operation for addition is subtraction and vice versa

> Inverse operation for division is multiplication

> Inverse operation for multiplication is division

11   www.math-knots.com

Inequality :

An inequality is a relation between two expressions that are not equal. As a mathematical statement an inequality states one side of the equation is less than, less than or equal to or greater than or greater than equal to the other side.

If the inequality has **less than** or **greater than** symbol,
1.   The graph starts with the open circle.
2.   For less than the graphing line goes toward the left.
3.   For greater than the graphing line goes toward the right.

If the inequality has **less than or equal to** or **greater than or equal** to symbol,
1.   The graph starts with the closed circle.
2.   For less than or equal to the graphing line goes toward the left.
3.   For greater than or equal to the graphing line goes toward the right.

Inequality statement	Inequality verbal expression	Inequality graph
x > -3 or -3 < x	x is greater than -3	-4 -3 -2 -1 0 1 2 3
x < 3 or 3 > x	x is less than 3	-3 -2 -1 0 1 2 3 4
x >= -1 or -1 <= x	x is greater than or equal to -1	-3 -2 -1 0 1 2 3
x <= 1 or 1 <= x	x is less than or equal to 1	-3 -2 -1 0 1 2 3

Basic inequalities :

Solving inequalities is same as solving for an equation except for one special rule.

www.math-knots.com

Compound Inequalities :

x < 0 or x ≥ 5 means all values less than 0 or 5 and more. In other words we are excluding the values 0,1,2,3,4.

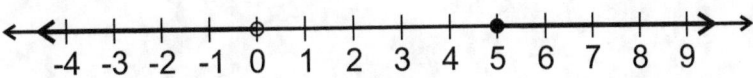

Absolute Value :

Absolute value of a number is its distance from 0. Since the distance cannot be negative absolute value is always positive.

$$|7| = 7 \qquad\qquad |-11| = 11$$

$$|2.3| = 2.3 \qquad\qquad |-0.75| = 0.75$$

Equations involving absolute values can be solved similar to regular algebraic equations solving. Absolute value should be treated as parentheses when applying PEDMAS rules.

Steps to solve absolute value equations.

Step 1 : Solve the expression within the absolute value.
(As applicable with PEDMAS rules)

Step 2 : Isolate the absolute value to one side of the equation.

Step 3 : Verify the value on the other side of the equation.
If the value is positive move to step 4.
If the value is negative there is no solution.

Step 4 : The expression inside the absolute value equals to positive and negative values of the other side of the equation.

Step 5 : Make the expression equal to positive value and solve for the variable.

Step 6 : Make the expression equal to negative value and solve for the variable.

Step 7 : The value obtained in step 5 and step 6 or the solutions to the absolute value equation.

Note : Absolute value equations can have two solutions. Since the absolute value of a number and its opposite are the same.
Absolute value can never be negative.

Absolute value Inequalities :

Absolute value inequalities are similar to absolute value equations.
Absolute value inequalities can have the below solutions
1. Two solutions
2. No solution
3. All real numbers

Steps to solve absolute value Inequalities are similar to solving the absolute value equations.

$|x| < a$ can be rewritten as $-a < x < a$ (where a is positive)
can also be written as $x < a$ **and** $x > -a$

$|x| \leq a$ can be rewritten as $-a \leq x \leq a$ (where a is positive)
can also be written as $x \leq a$ **and** $x \geq -a$

$|x| > a$ can be rewritten as $x > a$ **or** $x < -a$ (where a is positive)

Note : $<$ or $\leq$ are represented by the word **and**
$>$ or $\geq$ are represented by the word **or**

System of Linear equations :

Two or more linear equations can be solved by using elimination or substitution methods to find the common point or the points that satisfy both the equations.
The common points obtained are the solutions for the system of equations given.

Graphing System of Linear equations :

Two or more linear equations are graphed in the same way as graphing a linear equation.
The point at which the two lines intersect is the solution of the given pair of equations.
The solution can be cross verified by solving the equations as explained above.

Graphing System of Linear Inequalities :

Graphing two or more linear inequalities is similar to graphing a linear inequality and extending to more. The overlapping area of the inequalities graphed is the set of solutions that work for both system of linear inequalities graphed.

Graphing Absolute value of equations :

Graphing an absolute value of equation is similar to graphing an equation for y = absolute value of x consider all the possible order pair solutions to graph.
The point at which the graph turns to another direction is called the **vertex**.
The vertical line of symmetry for any graph is at its vertex.

Polynomials :

A polynomial is an expression or an equation with more than one term containing variables. A polynomial is named

      1. Its highest power of a variable
      Example : Linear (x), Quadratic (a^2), Cubic (b^3)

      2. By the number of terms of the polynomial
      Example : Monomial (x), Binomial (a^2 + b), Trinomial (a^3 + b^2 + c)

Polynomial is identified its two parts as described above. Polynomials must be simplified to the lowest possible terms before they are named.

Simplifying Standard form of polynomials :

Step 1 : Regroup the like terms and keep them in parentheses

Step 2 : Add or subtract the like terms as defined by signs preceeding the coefficients

Step 3 : Open the parentheses and make sure all the terms are simplified to the lowest possible.

Step 4 : Write the term with the highest followed by next lowest degree and so on.

This is the standard form of writing a polynomial

Example : $5x^3 + 3x - 6x^2 + 2x - 7x + x^3 + 9$

      $= (5x^3 + x^3) - 6x^2 + (3x + 2x - 7x) + 9$

      $= 6x^3 - 6x^2 - 2x + 9$

www.math-knots.com

Simplifying expressions :

Simplifying polynomial expressions is similar to simplifying any expressions.

1. Like terms can only be combined.
2. Distribute the numerical value or the variable as necessary or as given
3. Remember the unlike terms can never be combined together.
Example : We cannot add an x to an x^2, $2a^3$ to a^4, xy^2 to x^2y
4. Make sure all parentheses are simplified and opened.
5. Make sure everything that can be combined is combined.
6. Remember the sign before the term goes along with the term.
7. While using algebraic method when terms move from one side of the equation to the other side they change the sign before them to their opposite.

Solving proportions using equations :

Step 1 : Cross multiply the numerator of one side of the equation with the denominator in the opposite side also called as butterfly method.

Example : $\dfrac{x + 2}{3} \diagdown\diagup \dfrac{3x + 5}{5}$

$$5(x + 2) = 3(3x + 5)$$

Step 2 : Distribute where necessary

$$5(x + 10) = 3(3x + 5)$$

$$5x + 5(10) = 3(3x) + 3(5)$$
$$5x + 50 = 9x + 15$$

Step 3 : Bring all the like terms to one side of the equations, combine the like terms and solve using algebraic expressions method.

$$
\begin{array}{rcl}
5x + 50 &=& 9x + 15 \\
-15 && -15
\end{array}
$$
⟶ Solving by algebraic method

$$
\begin{array}{rcl}
5x + 35 &=& 9x \\
-5x && -5x
\end{array}
$$

$$35 = 9x - 5x$$ ⟶ Combining like terms
$$35 = 4x$$

$$\dfrac{4x}{4} = \dfrac{35}{4} = \boxed{x = \dfrac{35}{4}}$$

www.math-knots.com

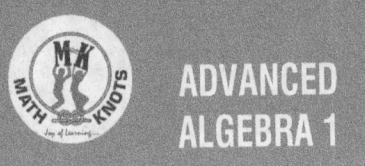

Multiplying binomials :

F.O.I.L acronym stands for "**Front, Outer, Inner, Last**"

F.O.I.L method is used to multiply binomials.
It can also be called as the **extended distributive property**

Lets multiply the binomials (ax + b) and (cx + d)

(ax + b) (cx + d)

Front : Multiplying the first terms of both binomials

(ax + b) (cx + d)

$ax.cx = acx^2$

Outer : Multiplying the first term of the first binomial with the second term (last term) of the second binomial

(ax + b) (cx + d)

ax.d = adx

Inner : Multiplying the second term of the first binomial with the first terms of the second binomial

(ax + b) (cx + d)

b.cx = bcx

Last : Multiplying the second terms (last terms) of both binomials

(ax + b) (cx + d)

b.d = bd

Write down all the terms obtained by using F.O.I.L method (Remember to consider the signs before the terms)
$acx^2 + adx + bcx + bd$

Combining the like terms, it can be simplified as $acx^2 + (ad + bc)x + bd$

Multiplying a binomial and a trinomial and more :

To multiply a binomial with a trinomial or a trinomial with a trinomial we use the extended distributive property.

Step 1 : Distribute the first term of the binomial or the trinomial into the trinomial

Step 2 : Distribute the second term of the binomial into the trinomial.

Step 3 : Repeat the above process for all the terms in the first trinomial or more.

Step 4 : Add all the products together obtained above. Remember to consider the signs in front of the terms while adding.

Step 5 : Combine the like terms. Remember to consider the signs in front of the terms while combining.

Step 6 : Make sure all the terms that can be combined are combined.

Step 7 : Write the polynomial in the standard simplified form.

Example : $(2x + 1)(x - 5)$

$(2x + 1)(x - 5)$ FRONT

$2x.x = 2x^2$

$(2x + 1)(x - 5)$ OUTER
$2x. -5 = -10x$

$(2x + 1)(x - 5)$ INNER
$1.x = x$

$(2x + 1)(x - 5)$ LAST
$1. -5 = -5$

Combining all the terms
$2x^2 + (-10x) + x + (-5)$
Combining like terms
$2x^2 -10x + x + -5$
$2x^2 -9x -5$

$(3x^2 + x - 2)(5x^2 - 3x + 7)$

$(3x^2)(5x^2 - 3x + 7) + (x)(5x^2 - 3x + 7) + (-2)(5x^2 - 3x + 7)$

$= 3x^2(5x^2 - 3x + 7) + x(5x^2 - 3x + 7) -2(5x^2 - 3x + 7)$

$= 3x^2(5x^2) + 3x^2(- 3x) + 3x^2 (7) + x(5x^2)$
$- x(3x) + x(7) -2(5x^2) - (-2)(3x) + (-2)(7)$

$= 15x^4 - 9x^3 + 21x^2 + 5x^3 - 3x^2 + 7x -10x^2 + 6x - 14$

Regroup the like terms

$= 15x^4 - 9x^3 + 5x^3 + 21x^2 - 3x^2 -10x^2 + 7x + 6x - 14$

$= 15x^4 - 4x^3 + 21x^2 - 13x^2 + 13x - 14$

$= 15x^4 - 4x^3 + 8x^2 + 13x - 14$

www.math-knots.com

Factoring a quadratic equation of the form $x^2 + bx + c$:

The coefficient of x^2 is 1 in this special case.

$(x + 5)(x - 3) = x^2 - 3x + 5x - 15 = x^2 + 2x - 15$

A quadratic equation $x^2 + 2x - 15$ can be factored by finding the factors for -15

Factors for -15 are : 1 & -15, -1 & 15, 3 & -5 and -3 & 5

Chose a pair of factors so that their sum is equal to +2

-3 and 5 are the right pair of factors as -3 + 5 = 2 and the remaining three pairs does not satisfy this criteria.

Rewrite $x^2 + 2x - 15 = x^2 - 3x + 5x - 15$

Take the greatest common factor x from the first two terms and 5 from the last two terms.

$x(x - 3) + 5(x - 3)$

Now x - 3 is the greatest common factor in both the terms

$(x - 3)(x + 5)$

$x^2 + 2x - 15 = (x - 3)(x + 5)$

Important : Remember to consider the sign before the terms when finding the common factors to factorize

---

Factoring a quadratic equation of the form $ax^2 + bx + c$:

Step 1 : Multiply the coefficient "a" and "c". Find the product.

Step 2 : Find all the factor products of "ac"

Step 3 : Identify a pair such that the product of two factors equals to "ac" and when the factors are added the sum equals to "b". Remember to consider the sign before the terms.

Step 4 : Rewrite the middle term as the sum of the identified factors.

Step 5 : Take the greatest common factor out of the first two terms and then from the last two terms.

Step 6 : If step 1 to 5 is done accurately we should see the matching binomial in the two terms.

Step 7 : Take the matching binomial out as the common factor.

Step 8 : After taking the matching binomial as a factor the coefficient of terms in step 6 become the second binomial.

Example : $2x^2 - 9x - 5$

2 X -5 = -10. Factors of -10 are 1 & -10, 10 & -1, 5 & -2, 2 & -5

Out of the factors 1 & -10 fit the criteria
1 + (-10) = -9 (Sum of the factors = -9)

10 & -1 does not fit the criteria.
10 + (-1) = 10 - 1 = 9 (We need the sum value as -9 not +9

$2x^2 - 9x - 5 = 2x^2 - 10x + x - 5$
$= 2x(x - 5) + 1(x - 5)$
$= (x - 5)(2x + 1)$

$2x^2 - 9x - 5 = (x - 5)(2x + 1)$

Solving a quadratic equation :

The quadratic equation $ax^2 + bx + c = 0$ can be solved by finding the factors and making teach of the factors equal to zero to find the values of x for which the quadratic equation becomes zero.

Example : $2x^2 - 9x - 5 = 0$

$(x - 5)(2x + 1) = 0$

$x - 5 = 0$ or $2x + 1 = 0$
$\quad + 5 \quad +5$ $\quad\quad - 1 \quad - 1$

$x = 5$ or $\dfrac{2x}{2} = \dfrac{-1}{2}$

$x = 5$ or $x = \dfrac{-1}{2}$

The quadratic function $2x^2 - 9x - 5 = 0$ when

$x = 5$ or $x = \dfrac{-1}{2}$

The solutions of the quadratic equation are

$5 , \dfrac{-1}{2}$

www.math-knots.com

Graphing quadratic functions :

An equation of the standard form $y = ax^2 + bx + c$ is a quadratic function that has an x^2

The graph of a quadratic function is called as a parabola. The parabola is always symmetrical at its vertex. The vertex is a point where the graph changes its direction.

1. If a is positive in the quadratic function $y = ax^2 + bx + c$ then the parabola opens up.

2. If a is negative in the quadratic function $y = ax^2 + bx + c$ the parabola opens down.

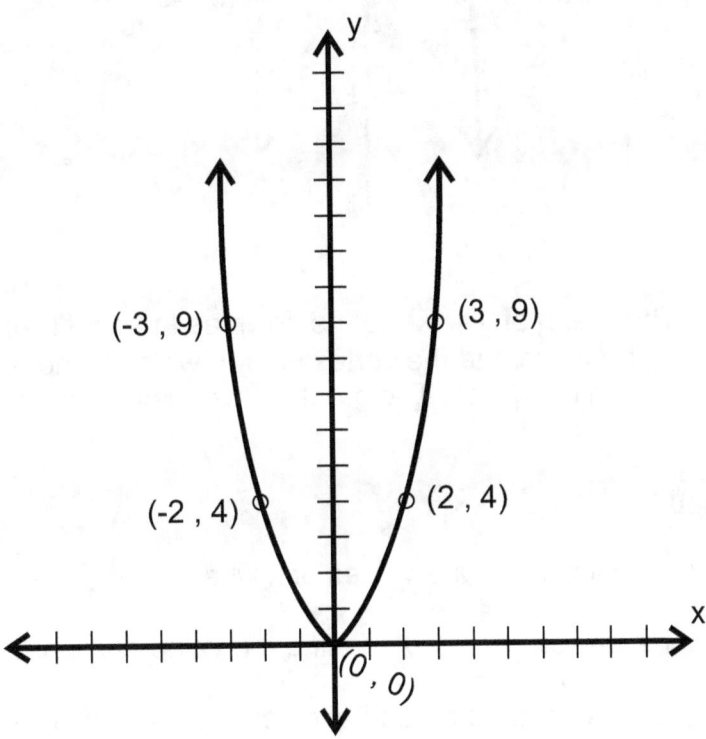

The graph of $y = x^2$ opens up as the coefficient of x^2
is positive and the vertex is (0,0) which is the lowest
point on the graph. The graph is symmetrical at the vertex.

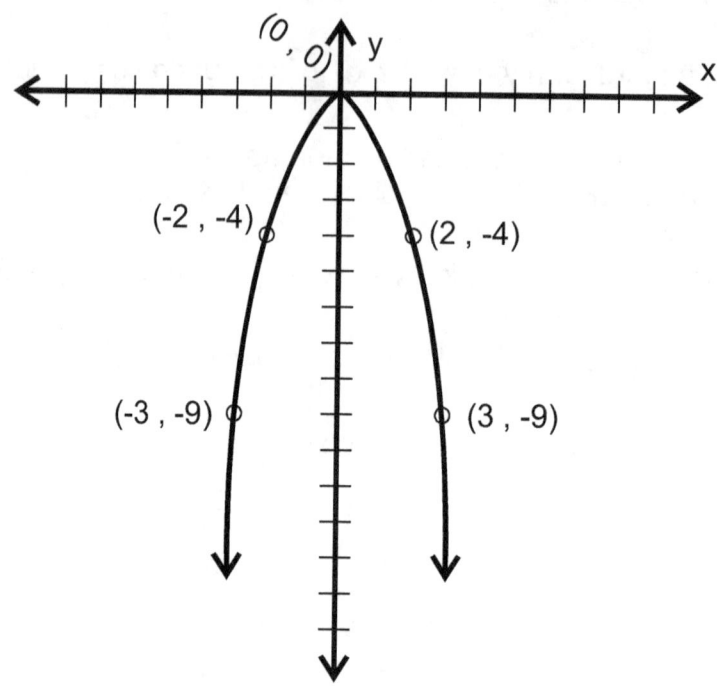

The graph of $y = -x^2$ opens down as the coefficient of x^2 is negative and the vertex is (0,0) which is the highest point on the graph. The graph is symmetrical at the vertex.

Finding x intercept :

The x intercept of any function are the values of x when y = 0.

1. For linear equations we can set y = 0 and find the value of x which is the x intercept.

2. For quadratic equations, factor the quadratic equation as explained in the previous sections and make it equal to 0 to find the values of x, which are x intercepts.

> x intercept is the point where the function value becomes 0.

Example : $y = 2x^2 - 9x - 5 = (x - 5)(2x + 1)$

To find x intercept make y = 0

so, $2x^2 - 9x - 5 = 0 = (x - 5)(2x + 1) = 0$

$(x - 5) = 0$, $(2x + 1) = 0$

$x = 5$ and $x = \frac{-1}{2}$

So the x intercepts are 5 and $\frac{-1}{2}$

If we plot a graph, the graph should touch at (5,0) and ($\frac{-1}{2}$, 0)

Finding y intercept :

The y intercept of any function are the values of y when x = 0.

1. For linear equations we can set x = 0 and find the value of y which is the y - intercept.

2. For quadratic equations, substitute x = 0 to find the values of y - intercept.

   y intercept is the point when x = 0.

Example : $y = 2x^2 - 9x - 5$

To find y intercept substitute x = 0
so, $2x^2 - 9x - 5 = y$
$y = 2(0^2) - 9(0) - 5$
$y = -5$

y - intercept is -5 for the quadratic equation $2x^2 - 9x - 5$.
The graph of the quadratic function $y = 2x^2 - 9x - 5$ touches the y axis at the point (0,-5)

Finding x intercept using quadratic formula :

The quadratic equation $y = ax^2 + bx + c = 0$ can also be factored by using the quadratic

formula $\dfrac{-b \pm \sqrt{b^2 - 4ac}}{2a}$

Step 1 : Make sure the quadratic equation is in the standard form $ax^2 + bx + c$ (convert it into standard form if it is not given)

Step 2 : If y is given then set y = 0.

Step 3 : Note down the values of "a" , "b" , "c". Make sure to consider the signs before each term.

Step 4 : Substitute the values "a" , "b" , "c" into the quadratic formula.

Step 5 : Since y = 0 the values found using the quadratic formula are the x - intercepts.

> It is not always possible to find the factors of the quadratic equation. In such a scenario the quadratic formula plays an important role. However we should get the same x - intercept values with both the methods.

The discriminant (b^2 - 4ac) :

The discriminant value (b^2 - 4ac) can be used to identified the number of x - intercepts for a given quadratic function.

1. If b^2 - 4ac = 0, the quadratic function will have one x - intercept which is $\frac{-b}{2a}$

2. If b^2 - 4ac is negative value and $\sqrt{b^2 - 4ac}$ is undefined (or imaginary), there are no x - intercepts for the given quadratic function.
In other words the graph will never touch the x axis.

3. If b^2 - 4ac is positive value, then the quadratic function will have two x - intercepts.

One intercept will be $\frac{-b + \sqrt{b^2 - 4ac}}{2a}$ and the other will be $\frac{-b - \sqrt{b^2 - 4ac}}{2a}$

Graphing a quadratic function :

The line of symmetry of any quadratic function of the standard form $y = ax^2 + bx + c = 0$ is at x = $\frac{-b}{2a}$

Step 1 : Point the line of symmetry by finding the value of $\frac{-b}{2a}$

Step 2 : Substitute the above value obtained in step 1 for x and find the y value of the vertex.

Step 3 : Find at least two other points to plot the graph.

Tip : You can substitute x = 0 to find the y - intercept value.

Step 4 : Plot the points and join them. You should see the graph as a parabola with the vertex and line of symmetry at $\frac{-b}{2a}$

Graphing a pair of quadratic functions :

Follow the same steps as explained above and graph each quadratic function.

Graphing a quadratic inequalities :

To graph a quadratic inequality perform the same steps as given for graphing the quadratic function and follow the below steps after plotting.

Step 5 : Draw a solid line for the quadratic function from step 4 if the inequality is $\geq$ or $\leq$ and move to step 7 otherwise move to step 6.

Step 6 : Draw a broken line for the quadratic function from step 4 if the inequality is $>$ or $<$ otherwise move to step 7.

Step 7 : Test one or two points outside the graph. If the test points satisfy the inequality meaning there are one of many possible solutions. Shade that side of the graph.

If the test points does not satisfy the inequality meaning there are not one of many possible solutions to the inequality. Shade the other side of the graph.

Graphing a pair of inequalities :

Follow the steps as explained in graphing the inequality and graph both the inequalities. Find the common area of the solutions and shade only that area. Any point lying in this common area is one of many solutions that satisfy the pair of quadratic inequalities.

25   www.math-knots.com

ADVANCED ALGEBRA 1
VOLUME 2

Identify the degree of the polynomial and also write the number of terms in each of the below.

(1) $-3a^2 + 6a + 9$

(2) $-3x^6 + 5x^4 - 2$

(3) -6

(4) $4v^8 + 4v^7 - v^6 - 9v^4 - 8v^3 - 7v^2$

(5) $x^8 - x^7 + 4x - 9$

(6) 6

(7) $-2x$

(8) $-2r^5 - 6$

(9) $-8k^7 - 10k^3$

(10) 9

Identify the degree of the polynomial and also write the number of terms in each of the below.

(11) -9

(12) $10x^3 + 7x$

(13) -10

(14) $7x^2$

(15) $9r^7$

(16) 5

(17) $-6n - 1$

(18) $-a^4 + 3a^3 + 3a^2 - 4$

(19) $-r^7$

(20) $-5x^3$

Identify the degree of the polynomial and also write the number of terms in each of the below.

(21) $3v^6$

(22) $8x^2 - 10x$

(23) $8p$

(24) $7a^3$

(25) $-6x^6 - 8x^5 + 8x^4 + 5x^2 - 10x + 7$

www.math-knots.com

Simplify each of the below expressions.

(26) $\left(x^2 - 4x^4\right) - \left(2x^4 - 7x\right)$

(27) $\left(5n^2 + n^4\right) - \left(2 + 5n^4\right)$

(28) $\left(6n^3 + 5\right) - \left(n^3 + 5\right)$

(29) $\left(4 - 3x\right) - \left(7 - 4x\right)$

(30) $\left(5n + 2n^2\right) - \left(7n^2 + 5n\right)$

(31) $\left(8x + x^4\right) - \left(6x^4 - 2x\right)$

(32) $\left(3n^4 + 6n^2\right) - \left(4n^3 + 8n^2\right)$

(33) $\left(6m^4 + 7m^2\right) - \left(3m^4 - 2m^2\right)$

(34) $\left(4b^3 + 3b^4\right) - \left(2b^4 + 3b^3\right)$

(35) $\left(1 - 3x^4\right) - \left(8 + 7x^4\right)$

Simplify each of the below expressions.

(36) $\left(2x^2 - 8x^4\right) + \left(4x^2 - 7x^4\right)$

(37) $\left(4 + 2x^3\right) + \left(6x^3 - 8\right)$

(38) $\left(5x - 3\right) + \left(1 + 2x\right)$

(39) $\left(4v^2 - 2\right) + \left(8 - 3v^2\right)$

(40) $\left(x + 4\right) + \left(6x - 8\right)$

(41) $\left(3k^4 - k^2\right) + \left(6k^4 - 8k^2\right)$

(42) $\left(5p^3 - 5\right) + \left(4 + 6p^3\right)$

(43) $\left(3p - 1\right) - \left(4 + 7p\right)$

(44) $\left(n^3 + 4n^2\right) - \left(6n^3 - 4n^2\right)$

(45) $\left(7x + 7x^4\right) - \left(6x - 2x^4\right)$

www.math-knots.com

Simplify each of the below expressions.

(46) $\left(8 - 4x^2\right) - \left(7x^2 + 7\right)$

(47) $\left(7x + 6\right) - \left(1 - x\right)$

(48) $\left(4b^4 + 5b\right) + \left(b - 4b^4\right)$

(49) $\left(3 + 2x^2\right) + \left(3x^2 + 1\right)$

(50) $\left(4v + 4v^4\right) - \left(3v^4 - 7v\right)$

(51) $\left(3 + b^2 - 2b^3\right) + \left(7b^3 - 3b^4 - 2b^2\right)$

(52) $\left(3 + a^2 - 8a\right) + \left(6a^2 + 4a + 5\right)$

(53) $\left(7n - 6n^3 - 7n^2\right) + \left(8n + 4 - n^3\right)$

(54) $\left(n^2 - 3 - n^3\right) + \left(3n^3 - 6n^2 + 7\right)$

(55) $\left(2x^3 - 5x^4 + 4x\right) - \left(x - 8 + 6x^3\right)$

www.math-knots.com

Simplify each of the below expressions.

(56) $\left(4x^2 + 4x^4 - 7x^3\right) + \left(6x^3 - 2x^2 - 5x^4\right)$

(57) $\left(p^4 - 7p^3 - 8\right) - \left(1 - 8p^3 + 4p^4\right)$

(58) $\left(3x + 8 - 6x^2\right) + \left(x^2 + 2x + 6\right)$

(59) $\left(5x^2 + 4 - 7x^3\right) + \left(7 + 8x^2 + 7x^3\right)$

(60) $\left(8p^2 + 3p - 8\right) + \left(6p^3 - 2 + 8p^2\right)$

(61) $\left(5b^3 - 7 - 2b\right) - \left(b - 7b^3 + 7b^2\right)$

(62) $\left(6 + a^4 + 7a^2\right) - \left(a^4 - 3 + a^2\right)$

(63) $\left(3b^3 + 5b - 4b^2\right) + \left(b + 2b^2 + 5b^3\right)$

(64) $\left(8x^4 - 1 + 8x^3\right) - \left(4x^3 + x^4 - 6\right)$

(65) $\left(6b^3 + 7b^2 + 6\right) - \left(6b^3 - 7b^4 - 7\right)$

Simplify each of the below expressions.

(66) $\left(2b^2 + 5b^3 - 4b^4\right) + \left(6b - 8b^3 - 4b^2\right)$

(67) $\left(2n^3 - 1 - 3n^2\right) - \left(5 - 7n^3 + 4n^2\right)$

(68) $\left(7n^2 - 2n^3 - 4n^4\right) + \left(3n^3 + n^2 + 4n^4\right)$

(69) $\left(x^2 - 5x - 8x^4\right) - \left(7 + 8x^2 + 6x^4\right)$

(70) $\left(3k^4 + 2k + 2k^2\right) + \left(2k - 7k^4 - 5k^2\right)$

(71) $\left(5 + 6b^3 + b\right) + \left(2b^3 - 4 + 6b\right)$

(72) $\left(3b + 4 - 7b^3\right) + \left(b + 5b^3 + 6\right)$

(73) $\left(m + 3m^4 - 6\right) - \left(3m + 4 - 7m^4\right)$

(74) $\left(6 + 8v^3 + 4v^4\right) - \left(3 + 3v^3 + 8v^4\right)$

(75) $\left(x + 5x^2 + 7\right) - \left(7 + 3x^2 - 4x\right)$

36   www.math-knots.com

Simplify each of the below expressions.

(76) $\left(3b - 4b^3 + 3b^4\right) + \left(5b^2 - 8b^3 - 6b^4\right) - \left(5b^3 - b^2 - 3b^4 - 3b\right)$

(77) $\left(5n^2 - 1 + 5n^3\right) + \left(5n^3 - 4n^4 + 1\right) + \left(5n^2 + 5n + 6 - 4n^3\right)$

(78) $\left(4k^4 + 5k + 2k^2\right) - \left(8k + 6k^2 + 5k^4\right) - \left(6k^3 - 2 - 5k - 7k^2\right)$

(79) $\left(5a^2 + 3a - 7a^3\right) - \left(5a + 8a^3 - 2a^2\right) + \left(7 + 4a^2 - 5a^3 - 7a\right)$

(80) $\left(2b^3 - 6b^2 - 2b\right) - \left(5 - 2b^4 + 4b^3\right) + \left(8b^2 + 2b^4 - 4b^3 + 5\right)$

Simplify each of the below expressions.

(81) $\left(5x^3 + 7 + x^4\right) - \left(7x^2 + x^3 - 3\right) + \left(2 - x^4 + 3x^3 - 3x^2\right)$

(82) $\left(2m^3 - 6 - 3m^4\right) - \left(6m + 8m^3 + 5\right) - \left(6m + 8m^3 - 6m^4 + 2\right)$

(83) $\left(3k^2 + 4k^4 + k\right) - \left(8k^4 - 5k - 2k^3\right) + \left(k^3 - 5k^4 + 7k^2 - k\right)$

(84) $\left(3 + 7n + n^4\right) + \left(5n^4 + 2 + 3n^2\right) - \left(2n^3 + 7 + 4n - n^4\right)$

(85) $\left(1 + 5b + 6b^2\right) + \left(2b^4 + 3 + 2b\right) - \left(5b^2 + 4 - 2b^4 + 6b\right)$

Simplify each of the below expressions.

(86) $\left(8m^3 - 7m + 1\right) - \left(7m^3 + 1 + 3m^4\right) + \left(3m^4 + 8m - 4m^3 - 3\right)$

(87) $\left(1 + 3x^4 - x^2\right) + \left(5 + x^3 + 2x^2\right) - \left(7 - 6x^4 + 7x^3 - 6x^2\right)$

(88) $\left(3b^3 - 6 - 8b^2\right) + \left(8 + 5b^4 - b^2\right) - \left(6b + b^2 + 5b^3 - 2\right)$

(89) $\left(3x^3 + 4x^2 + 4\right) + \left(1 - 4x^2 + x\right) + \left(7x + 3x^2 - 4 - 5x^3\right)$

(90) $\left(8r^3 + 4r + 6r^4\right) - \left(8r^4 + 3 + 6r^3\right) + \left(8 + 6r - 3r^4 - 7r^3\right)$

Simplify each of the below expressions.

(91) $\left(4 + 2n^3 - 5n^4\right) + \left(4n + 2n^4 - n^2\right) + \left(6n^4 - n^2 + 8n + 3n^3\right)$

(92) $\left(3n^2 - 5n^3 + 2\right) + \left(5n + 7 - 6n^3\right) - \left(3n + n^3 + 7n^4 - 2\right)$

(93) $\left(n^3 + 3n^4 + 8n\right) + \left(3n^4 - 4n^3 - 6n\right) + \left(4n^4 + 5n - 7n^2 + 8n^3\right)$

(94) $\left(8x - 2 - 7x^3\right) - \left(x - 2x^3 - 3x^4\right) - \left(4x^3 + 4 - 4x + 5x^4\right)$

(95) $\left(6x^3 + 4x^4 - 2\right) + \left(x^2 + x^4 - 7x^3\right) - \left(2x - 3x^3 + 3x^2 + 4x^4\right)$

Simplify each of the below expressions.

(96) $\left(4a^2 - 4a^3 - 6a\right) - \left(6a^2 + 4a - 2a^4\right) - \left(2a - 7a^4 - a^2 + 2a^3\right)$

(97) $\left(6 - 4x + 2x^3\right) + \left(4 - 6x^4 - 7x^2\right) + \left(5x + 7 - 7x^3 - x^4\right)$

(98) $\left(7 + n^4 + 2n^2\right) + \left(5 - 5n - 6n^4\right) - \left(8n + 4 + 6n^2 + 8n^4\right)$

(99) $\left(8p + 2 + p^3\right) + \left(3 - 5p^3 - 4p^2\right) - \left(p^2 + 5p^3 - 6 - 6p\right)$

(100) $\left(5n + 2n^2 + 2\right) - \left(6n^2 - 7 - 8n^3\right) - \left(3n + 6n^2 + 8n^3 + 1\right)$

For more visit www.a4ace.com   41   www.math-knots.com

Simplify each of the below expressions.

(101) $\left(2x^4 + 3x + 6x^3\right) + \left(8 - 2x - 8x^3\right) - \left(3 - x^4 + 3x + 8x^3\right)$

(102) $\left(3r^3 + r + 3r^2\right) - \left(6r^2 + 8r^4 - 2r^3\right) + \left(3r^2 - 5r^4 + 8 + 6r^3\right)$

(103) $\left(8k - 6k^2 + 4k^4\right) + \left(8 - 5k^3 - 8k^2\right) - \left(4 - k + 8k^4 + k^3\right)$

(104) $\left(6n^3 + 4n^4 + 3n\right) + \left(n^4 + 7 - 5n^2\right) - \left(8n^2 - 4n^3 + 8n^4 + n\right)$

(105) $\left(3 - 6x - 2x^4\right) - \left(2 + 5x^3 + 2x\right) - \left(2 + x^3 + 7x^4 + 6x\right)$

Divide each of the below given polynomial by its divisor and find the quotient.

(106) $\left(40n^3 + 2n^2 + 4n\right) \div 10\,n^3$

(107) $\left(2k^4 + 2k^3 + 3k^2\right) \div 6\,k^2$

(108) $\left(2x^3 + 3x^2 + 3x\right) \div 8x^2$

(109) $\left(4r^4 + 50r^3 + 5r^2\right) \div 10\,r^2$

(110) $\left(2k^3 + 3k^2 + 18k\right) \div 6\,k^2$

(111) $\left(4k^4 + 3k^3 + 30k^2\right) \div 6\,k^3$

(112) $\left(5k^3 + 5k^2 + 10k\right) \div 10k$

(113) $\left(3x^3 + 2x^2 + 5x\right) \div 6x$

(114) $\left(50k^3 + 50k^2 + 2k\right) \div 10k^3$

(115) $\left(12x^3 + 6x^2 + x\right) \div 6x^3$

43   www.math-knots.com

Divide each of the below given polynomial by its divisor and find the quotient.

(116) $\left(2n^4 + 9n^3 + 18n^2\right) \div 9n$

(117) $\left(2a^3 + 6a^2 + 18a\right) \div 6a^2$

(118) $\left(40p^7 + 3p^6 + 16p^5\right) \div 8p^2$

(119) $\left(40x^3 + 30x^2 + 4x\right) \div 10x$

(120) $\left(24n^4 + n^3 + 4n^2\right) \div 8n^3$

(121) $\left(2n^4 + 40n^3 + 8n^2\right) \div 8n^3$

(122) $\left(24n^3 + 2n^2 + 6n\right) \div 6n^2$

(123) $\left(12v^6 + 18v^5 + 30v^4\right) \div 6v$

(124) $\left(4k^3 + k^2 + 4k\right) \div 4k$

(125) $\left(24v^3 + 2v^2 + 12v\right) \div 6v^3$

44   www.math-knots.com

Divide each of the below given polynomial by its divisor and find the quotient.

(126) $\left(2x^3 + 2x^2 + 12x\right) \div 6x$

(127) $\left(50x^3 + 2x^2 + 2x\right) \div 10x^3$

(128) $\left(3x^6 + 30x^5 + x^4\right) \div 6x$

(129) $\left(27n^3 + 45n^2 + 9n\right) \div 9n^3$

(130) $\left(8m^3 + 8m^2 + 8m\right) \div 8m^2$

(131) $\left(m^2 - 4m - 15\right) \div \left(m - 7\right)$

(132) $\left(n^2 - 4n - 62\right) \div \left(n + 6\right)$

(133) $\left(b^2 + b - 99\right) \div \left(b + 10\right)$

(134) $\left(k^2 + 7k - 40\right) \div \left(k + 10\right)$

(135) $\left(v^2 - 7v - 26\right) \div \left(v - 9\right)$

Divide each of the below given polynomial by its divisor and find the quotient.

(136) $\left(m^2 + 8m - 11\right) \div (m - 2)$

(137) $\left(b^2 - 5b - 76\right) \div (b + 7)$

(138) $\left(7k^2 + 62k - 5\right) \div (k + 9)$

(139) $\left(x^2 + 5x - 5\right) \div (x - 1)$

(140) $\left(n^2 + 21n + 102\right) \div (n + 11)$

(141) $\left(7r^2 - 52r + 19\right) \div (r - 7)$

(142) $\left(2a^2 - 33a + 117\right) \div (a - 12)$

(143) $\left(x^2 - 9x - 37\right) \div (x + 3)$

(144) $\left(n^2 + 15n + 60\right) \div (n + 8)$

(145) $\left(r^2 + 8r + 3\right) \div (r + 9)$

Divide each of the below given polynomial by its divisor and find the quotient.

(146) $\left(x^2 - 4x - 107\right) \div \left(x + 8\right)$

(147) $\left(x^2 + 12x + 40\right) \div \left(x + 4\right)$

(148) $\left(x^2 - x - 9\right) \div \left(x + 4\right)$

(149) $\left(n^2 + 3n - 16\right) \div \left(n + 5\right)$

(150) $\left(n^2 - 14n + 27\right) \div \left(n - 3\right)$

(151) $\left(x^2 + 10x + 1\right) \div \left(x + 11\right)$

(152) $\left(r^2 + 17r + 69\right) \div \left(r + 10\right)$

(153) $\left(r^2 + r - 34\right) \div \left(r - 5\right)$

(154) $\left(k^2 + 3k - 11\right) \div \left(k - 3\right)$

(155) $\left(x^2 + 15x + 51\right) \div \left(x + 11\right)$

www.math-knots.com

Divide each of the below given polynomial by its divisor and find the quotient.

(156) $\left(n^3 - 4n^2 - 29n + 56\right) \div (n - 7)$

(157) $\left(6x^3 + 64x^2 + 31x - 90\right) \div (x + 10)$

(158) $\left(5a^3 + 13a^2 - 15a - 27\right) \div (a + 3)$

(159) $\left(k^3 + 10k^2 + 28k + 16\right) \div (k + 4)$

(160) $\left(n^3 - 2n^2 - 67n + 36\right) \div (n - 9)$

(161) $\left(v^3 - 16v^2 + 64v - 9\right) \div (v - 9)$

(162) $\left(4v^3 - 26v^2 - 9v - 35\right) \div (v - 7)$

(163) $\left(r^3 + 6r^2 + 14r + 15\right) \div (r + 3)$

(164) $\left(m^3 + 7m^2 - 33m + 9\right) \div (m - 3)$

(165) $\left(n^3 + 7n^2 - 34n + 12\right) \div (n - 3)$

Divide each of the below given polynomial by its divisor and find the quotient.

(166) $\left(p^3 - 14p^2 + 36p + 16\right) \div \left(p - 4\right)$

(167) $\left(k^3 + 2k^2 - 10k + 25\right) \div \left(k + 5\right)$

(168) $\left(x^3 - 9x^2 + 9x + 54\right) \div \left(x - 6\right)$

(169) $\left(x^3 + x^2 - 58x + 14\right) \div \left(x - 7\right)$

(170) $\left(2x^3 + 3x^2 - 36x - 5\right) \div \left(x + 5\right)$

(171) $\left(x^3 - 4x^2 - 52x - 35\right) \div \left(x + 5\right)$

(172) $\left(10p^3 - 42p^2 + 7p + 4\right) \div \left(p - 4\right)$

(173) $\left(3m^3 + 16m^2 + 22m + 4\right) \div \left(m + 2\right)$

(174) $\left(k^3 + 3k^2 - 31k + 72\right) \div \left(k + 8\right)$

(175) $\left(n^3 - 15n^2 + 60n - 36\right) \div \left(n - 6\right)$

49   www.math-knots.com

Divide each of the below given polynomial by its divisor and find the quotient.

(176) $\left(4x^3 - 4x^2 - 16x - 24\right) \div (x - 3)$

(177) $\left(m^3 + 20m^2 + 97m - 30\right) \div (m + 10)$

(178) $\left(m^3 + 4m^2 - 39m - 30\right) \div (m - 5)$

(179) $\left(r^3 + 14r^2 + 45r - 24\right) \div (r + 8)$

(180) $\left(b^3 - b^2 - 16b + 16\right) \div (b - 4)$

Multiply each of the given expressions to its simplified form.

(181) $(7p + 2)(3p - 8)$

(182) $(3a - 1)(8a - 7)$

(183) $(7x - 2)(x - 4)$

(184) $(7b - 4)(7b - 8)$

(185) $(4a - 1)(5a + 4)$

(186) $(p + 5)(3p - 1)$

(187) $(5k + 1)(k - 2)$

(188) $(x + 1)(2x - 8)$

(189) $(4n + 3)(2n - 4)$

(190) $(4x - 4)(5x - 1)$

Multiply each of the given expressions to its simplified form.

(191) $(8k - 3)(2k + 4)$

(192) $(2m - 8)(m + 6)$

(193) $(2p + 6)(2p - 6)$

(194) $(2v + 2)(7v + 7)$

(195) $(5x + 6)(5x + 4)$

(196) $(5r + 7)(4r + 4)$

(197) $(n + 8)(3n - 6)$

(198) $(5k + 8)(5k - 4)$

(199) $(a - 2)(3a + 3)$

(200) $(4x - 4)(8x - 6)$

Multiply each of the given expressions to its simplified form.

(201) $(2a - 4)(6a - 2)$

(202) $(2n + 5)(6n + 8)$

(203) $(4x + 2)(6x - 3)$

(204) $(2a + 3)(4a - 4)$

(205) $(6x + 1)(6x + 4)$

(206) $(3n + 6)(5n^2 + 5n + 7)$

(207) $(n + 3)(2n^2 - n + 4)$

(208) $(5x + 3)(x^2 - x + 1)$

(209) $(6x - 5)(8x^2 - 4x + 5)$

(210) $(4x + 2)(x^2 + 3x + 8)$

Multiply each of the given expressions to its simplified form.

(211) $(2m - 1)(6m^2 - 3m - 6)$

(212) $(5a + 1)(6a^2 - 4a - 4)$

(213) $(7x - 7)(5x^2 - 7x + 6)$

(214) $(2a + 1)(2a^2 + 3a - 1)$

(215) $(6k + 7)(3k^2 + 5k - 6)$

(216) $(3n + 3)(6n^2 - 7n - 1)$

(217) $(5v - 1)(v^2 + 2v - 1)$

(218) $(x + 7)(3x^2 + x + 7)$

(219) $(6n - 8)(3n^2 + 2n - 6)$

(220) $(2n - 6)(n^2 + 7n + 3)$

54   www.math-knots.com

Multiply each of the given expressions to its simplified form.

(221) $(n+2)(8n^2 - n + 3)$

(222) $(4k-1)(5k^2 + k - 3)$

(223) $(8n+1)(n^2 - 8n - 6)$

(224) $(n-6)(8n^2 + 5n - 1)$

(225) $(2k+1)(6k^2 - 4k + 6)$

(226) $(8m-7)(m^2 + 5m - 3)$

(227) $(8b-7)(6b^2 + 3b + 4)$

(228) $(n-2)(3n^2 + 6n + 8)$

(229) $(4p-3)(3p^2 - 3p - 1)$

(230) $(4k+3)(5k^2 - 6k - 5)$

Multiply each of the given expressions to its simplified form.

(231) $(7r + 6)(r^2 - 3r - 5)$

(232) $(8n + 6)(6n^2 - n + 8)$

(233) $(2x - 7)(3x^2 + 6x + 7)$

(234) $(3n - 8)(4n^2 - 4n - 1)$

(235) $(5a + 7)(a^2 + a + 8)$

(236) $(n^2 + 4n + 2)(n^2 - 8n - 2)$

(237) $(3x^2 - 4x + 1)(5x^2 - x + 1)$

(238) $(5m^2 + m - 2)(3m^2 - 3m + 3)$

(239) $(5m^2 - m - 5)(m^2 - 3m + 7)$

(240) $(6k^2 + 5k - 1)(4k^2 - 8k + 5)$

www.math-knots.com

Multiply each of the given expressions to its simplified form.

(241) $\left(2k^2 - 2k + 2\right)\left(7k^2 + 3k - 5\right)$

(242) $\left(3n^2 - 4n - 5\right)\left(n^2 + 2n + 2\right)$

(243) $\left(6r^2 + 2r - 3\right)\left(7r^2 + 4r - 2\right)$

(244) $\left(6x^2 + 8x + 7\right)\left(3x^2 - x + 1\right)$

(245) $\left(x^2 + 3x - 7\right)\left(7x^2 - 5x + 7\right)$

(246) $\left(7v^2 + v - 8\right)\left(6v^2 - 3v - 5\right)$

(247) $\left(4x^2 - 8x - 7\right)\left(6x^2 - 2x + 7\right)$

(248) $\left(3x^2 + 3x + 8\right)\left(4x^2 + 2x - 5\right)$

(249) $\left(8k^2 + k + 1\right)\left(3k^2 - 7k + 2\right)$

(250) $\left(7v^2 + 4v - 8\right)\left(4v^2 - 4v - 6\right)$

57   www.math-knots.com

Multiply each of the given expressions to its simplified form.

(251) $\left(x^2 - x + 4\right)\left(6x^2 - 2x + 4\right)$

(252) $\left(7m^2 - m - 8\right)\left(6m^2 - 6m - 6\right)$

(253) $\left(6x^2 - 7x + 2\right)\left(4x^2 + 2x + 7\right)$

(254) $\left(7n^2 - 7n - 4\right)\left(2n^2 - 3n - 7\right)$

(255) $\left(7x^2 - 8x - 3\right)\left(8x^2 + 8x + 2\right)$

(256) $\left(3r^2 + 8r - 2\right)\left(3r^2 - 3r + 6\right)$

(257) $\left(8k^2 - 6k + 1\right)\left(5k^2 + 3k - 5\right)$

(258) $\left(8x^2 - x + 1\right)\left(3x^2 - 3x + 6\right)$

(259) $\left(4v^2 + 6v - 8\right)\left(6v^2 - 6v + 1\right)$

(260) $\left(7n^2 - 2n + 1\right)\left(n^2 + 4n - 7\right)$

Find the common factor and find rewrite the given expressions as a product of its factors.

(261) $45a^2 + 50a$

(262) $-9a^2 - 3a^3$

(263) $14x^4 + 21x^2$

(264) $-32b^2 - 36b$

(265) $-50x + 20$

(266) $42a^5 + 24a$

(267) $18x + 48$

(268) $-35x^3 + 14x^2$

(269) $-3 + 9a^2$

(270) $-64b + 24$

59

www.math-knots.com

Find the common factor and find rewrite the given expressions as a product of its factors.

(271) $9k^8 + 18k$

(272) $8m^6 + 36m^5$

(273) $12a^5 - 18a^4$

(274) $-21p^4 + 14p^3$

(275) $5k^4 - 40$

(276) $-56x^5 + 48x^2$

(277) $-45r^2 + 36r$

(278) $8x^3 - 20x$

(279) $10p^4 + 8p^2$

(280) $-5b^5 + 10b^2$

Find the common factor and find rewrite the given expressions as a product of its factors.

(281) $24v^7 - 64$

(282) $28x^2 - 35$

(283) $63m^3 + 9$

(284) $100r^2 - 70r^3$

(285) $2m^4 + 4$

Find the product of the below.

(286) $(x+3y)^2$

(287) $(a-14b)^2$

(288) $(n+45m)^2$

(289) $(y-34x)^2$

(290) $(u-9v)^2$

(291) $(x+31y)^2$

(292) $(a-25b)^2$

(293) $(u+41v)^2$

(294) $(x+39y)^2$

(295) $(y+22x)^2$

Find the product of the below.

(296) $(v - 29u)^2$

(297) $(u - 42v)^2$

(298) $(a - 2b)^2$

(299) $(u + 23v)^2$

(300) $(x + 47y)^2$

(301) $(y + 46x)^2$

(302) $(n - 28m)^2$

(303) $(a - 31b)^2$

(304) $(m + 6n)^2$

(305) $(x - 3y)^2$

www.math-knots.com

Find the product of the below.

(306) $(n + 2m)^2$

(307) $(a + 50b)^2$

(308) $(a + 49b)^2$

(309) $(x + 33y)^2$

(310) $(v - 26u)^2$

(311) $(m + 14n)^2$

(312) $(u + 48v)^2$

(313) $(u - 20v)^2$

(314) $(x - 43y)^2$

(315) $(y - 11x)^2$

Find the product of the below.

(316) $(x - 10y)(x + 10y)$

(317) $(x - 7y)(x + 7y)$

(318) $(x + 6y)(x - 6y)$

(319) $(x + 12y)(x - 12y)$

(320) $(m + 14n)(m - 14n)$

(321) $(u - 11v)(u + 11v)$

(322) $(u - 9v)(u + 9v)$

(323) $(m + 13n)(m - 13n)$

(324) $(b + 11a)(b - 11a)$

(325) $(x + 3y)(x - 3y)$

www.math-knots.com

Find the product of the below.

(326) $(x + 5y)(x - 5y)$

(327) $(y + 4x)(y - 4x)$

(328) $(u + 2v)(u - 2v)$

(329) $(n - 4m)(n + 4m)$

(330) $(x + 10y)(x - 10y)$

(331) $(x + 9y)(x - 9y)$

(332) $(m - n)(m + n)$

(333) $(x - 13y)(x + 13y)$

(334) $(b - 12a)(b + 12a)$

(335) $(u - 5v)(u + 5v)$

66   www.math-knots.com

Find the product of the below.

(336) $(y + 7x)(y - 7x)$

(337) $(x + y)(x - y)$

(338) $(u + 8v)(u - 8v)$

(339) $(v - 6u)(v + 6u)$

(340) $(u - 3v)(u + 3v)$

67

www.math-knots.com

Factorize the below expressions to the lowest possible.

(341) $84x^2 + 28xy - 231y^2$

(342) $12u^2 + 41uv + 24v^2$

(343) $72x^2 + 584xy + 560y^2$

(344) $14x^2 + 26xy$

(345) $9x^2 - 16xy - 4y^2$

(346) $40x^2 - 405xy - 385y^2$

(347) $8x^2 + 75xy + 88y^2$

(348) $10a^2 + 71ab + 7b^2$

(349) $14x^2 + 3xy - 36y^2$

(350) $10u^2 - 3uv - v^2$

68   www.math-knots.com

Factorize the below expressions to the lowest possible.

(351) $8x^2 - 43xy - 30y^2$

(352) $14a^2 + 107ab - 40b^2$

(353) $9a^2 + 32ab - 16b^2$

(354) $36x^2 + 204xy + 280y^2$

(355) $16a^2 - 172ab + 352b^2$

(356) $9x^2 - xy - 10y^2$

(357) $20x^2 - 42xy + 18y^2$

(358) $12x^2 - 118xy - 20y^2$

(359) $12u^2 + 74uv - 176v^2$

(360) $4x^2 - 13xy + 3y^2$

ADVANCED ALGEBRA 1

Factorize the below expressions to the lowest possible.

(361) $8x^2 - 26xy + 15y^2$

(362) $9a^2 + 6ab - 80b^2$

(363) $12x^2 + 8xy - 7y^2$

(364) $9x^2 + 28xy + 20y^2$

(365) $4x^2 + 19xy + 22y^2$

(366) $10u^2 - 61uv + 33v^2$

(367) $36x^2 - 36xy - 280y^2$

(368) $28x^2 - 2xy - 60y^2$

(369) $6m^2 + 16mn$

(370) $45x^2 + 430xy - 200y^2$

Factorize the below expressions to the lowest possible.

(371) $9n^2 - 77n - 36$

(372) $6m^2 - 5m - 99$

(373) $70x^2 - 133x + 63$

(374) $9x^2 - 82x + 80$

(375) $12x^2 - 38x + 16$

(376) $8n^2 - 63n - 81$

(377) $9k^2 + 38k + 8$

(378) $10p^2 - 39p + 27$

(379) $45x^2 - 440x - 100$

(380) $10v^2 + 99v + 81$

71

www.math-knots.com

Factorize the below expressions to the lowest possible.

(381) $10n^2 + 59n + 84$

(382) $4k^2 + 43k + 30$

(383) $12x^2 + 55x + 28$

(384) $12v^2 + 129v + 90$

(385) $6n^2 - n - 5$

(386) $4n^2 + 31n - 90$

(387) $12a^2 - 98a - 90$

(388) $9a^2 - 92a + 20$

(389) $6n^2 + n - 2$

(390) $54b^2 + 102b - 12$

72
www.math-knots.com

Factorize the below expressions to the lowest possible.

(391) $54b^2 + 306b + 420$

(392) $50n^2 - 15n - 90$

(393) $20n^2 + 114n + 108$

(394) $24r^2 - 140r + 200$

(395) $60p^2 - 306p - 324$

(396) $36x^2 + 42x - 120$

(397) $16p^2 + 4p - 42$

(398) $45r^2 + 185r + 20$

(399) $12m^2 + 45m - 12$

4(00) $27x^2 - 39x + 12$

Factorize the below expressions to the lowest possible.

(401) $18a^2 + 69a + 45$

(402) $12x^2 + 14x - 180$

(403) $12n^2 - 46n + 20$

(404) $54n^2 - 6n - 48$

(405) $27v^2 + 150v - 72$

(406) $12x^2 - 38x - 72$

(407) $40x^2 + 148x - 144$

(408) $48v^2 - 126v + 60$

(409) $18n^2 - 172n - 80$

(410) $30p^2 - 85p - 50$

74   www.math-knots.com

Factorize the below expressions to the lowest possible.

(411) $4u^2 - 36v^2$

(412) $18m^2 - 32n^2$

(413) $18u^2 - 8v^2$

(414) $36x^2 - 64y^2$

(415) $80x^2 - 125y^2$

(416) $100x^2 - 16y^2$

(417) $4a^2 - 64b^2$

(418) $27x^2 - 75y^2$

(419) $100x^2 - 4y^2$

(420) $4x^2 - 100y^2$

www.math-knots.com

Factorize the below expressions to the lowest possible.

(421) $3x^2 - 12y^2$

(422) $8x^2 - 2y^2$

(423) $5x^2 - 20y^2$

(424) $75x^2 - 12y^2$

(425) $45x^2 - 5y^2$

(426) $75x^2 - 48y^2$

(427) $64u^2 - 100v^2$

(428) $5x^2 - 5y^2$

(429) $5u^2 - 125v^2$

(430) $2x^2 - 32y^2$

Factorize the below expressions to the lowest possible.

(431) $588x^2 - 1848xy + 1452y^2$

(432) $980x^2 - 140xy + 5y^2$

(433) $256x^2 - 576xy + 324y^2$

(434) $2197x^2 - 1690xy + 325y^2$

(435) $891x^2 + 1980xy + 1100y^2$

(436) $176x^2 - 88xy + 11y^2$

(437) $432x^2 - 1872xy + 2028y^2$

(438) $588x^2 + 504xy + 108y^2$

(439) $3x^2 - 42xy + 147y^2$

(440) $99x^2 + 132xy + 44y^2$

Factorize the below expressions to the lowest possible.

(441) $12u^2 - 72uv + 108v^2$

(442) $176u^2 + 792uv + 891v^2$

(443) $325x^2 + 1820xy + 2548y^2$

(444) $325x^2 + 1560xy + 1872y^2$

(445) $900x^2 - 540xy + 81y^2$

(446) $567x^2 + 1008xy + 448y^2$

(447) $11x^2 - 154xy + 539y^2$

(448) $192a^2 + 288ab + 108b^2$

(449) $98x^2 - 140xy + 50y^2$

(450) $2197x^2 + 3380xy + 1300y^2$

Factorize the below expressions to the lowest possible.

(451) $900x^2 + 2340xy + 1521y^2$

(452) $75a^2 - 180ab + 108b^2$

(453) $1183u^2 - 1274uv + 343v^2$

(454) $225x^2 - 360xy + 144y^2$

(455) $2016x^2 + 2352xy + 686y^2$

(456) $56bz + 49bc - 40xz - 35xc$

(457) $10xc + xf + 60yc + 6yf$

(458) $90bz^2 - 50bh - 63xz^2 + 35xh$

(459) $56az^2 + 7ah - 48yz^2 - 6yh$

(460) $2xu^2 + 3xv - 2yu^2 - 3yv$

79   www.math-knots.com

Factorize the below expressions to the lowest possible.

(461) $10x^2u + 5x^2v + 8yu + 4yv$

(462) $14xw + 35xf - 6yw - 15yf$

(463) $70bz^2 - 49bh + 60xz^2 - 42xh$

(464) $12xc - 16xd + 3yc - 4yd$

(465) $63bc^2 - 7bf^2 - 18xc^2 + 2xf^2$

(466) $81ac + 90af + 72bc + 80bf$

(467) $15xc - 20xf + 21yc - 28yf$

(468) $7ac + 8af - 63bc - 72bf$

(469) $63ah^2 - 56af - 90yh^2 + 80yf$

(470) $27pu + 21pv - 63qu - 49qv$

www.math-knots.com

Factorize the below expressions to the lowest possible.

(471) $70ah - 20af + 63yh - 18yf$

(472) $3mh + 6mf + 7nh + 14nf$

(473) $70xc + 63xf + 60yc + 54yf$

(474) $24xz + 64xh - 27yz - 72yh$

(475) $5bz - 15bh - 3xz + 9xh$

(476) $3au - 21av - xu + 7xv$

(477) $28ac - 35ad + 24yc - 30yd$

(478) $9p^2c - 24p^2d^2 - 30qc + 80qd^2$

(479) $56mh + 7mf + 8nh + nf$

(480) $8a^2z - 20a^2c - 2xz + 5xc$

Volume 2

Factorize the below expressions to the lowest possible.

(481) $50ah + 35af - 10bh - 7bf$

(482) $5a^2z + 4a^2h + 15yz + 12yh$

(483) $7bz - 6bh + 7xz - 6xh$

(484) $90ah + 27af - 100bh - 30bf$

(485) $10bh + 10bf + 9xh + 9xf$

www.math-knots.com

Plot the graph for each of the given functions below.

(486) $y = \dfrac{1}{2}(x+4)^2 - 4$

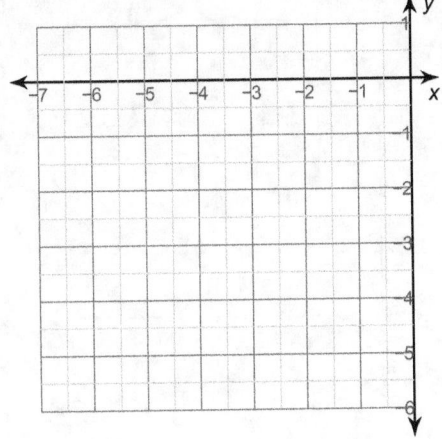

(487) $y = (x+2)^2 + 2$

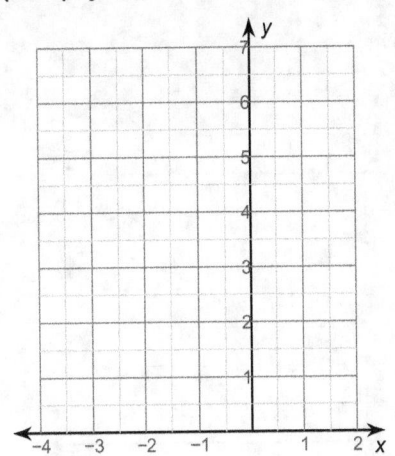

(488) $y = 3(x-2)^2 + 2$

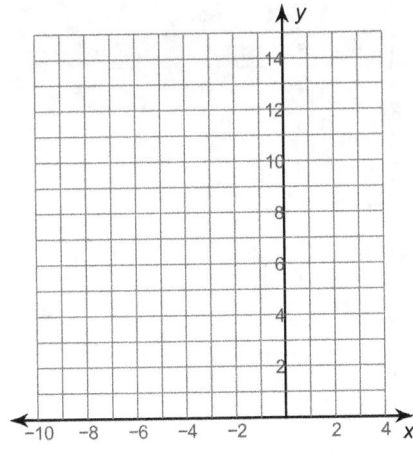

(489) $y = \dfrac{1}{2}(x+2)^2 - 3$

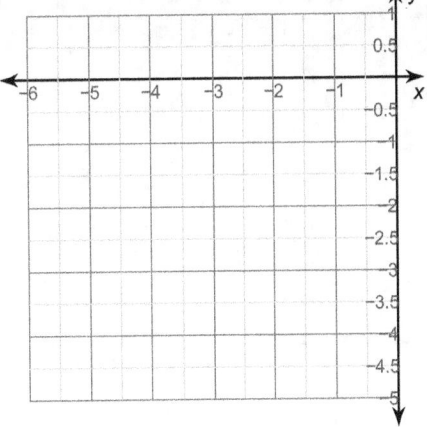

83

Plot the graph for each of the given functions below.

(490) $y = 2(x - 3)^2 + 1$

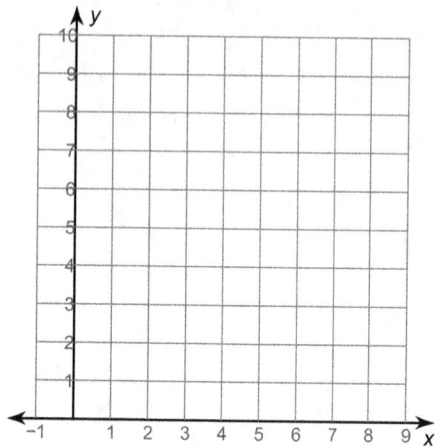

(491) $y = -(x - 3)^2 + 2$

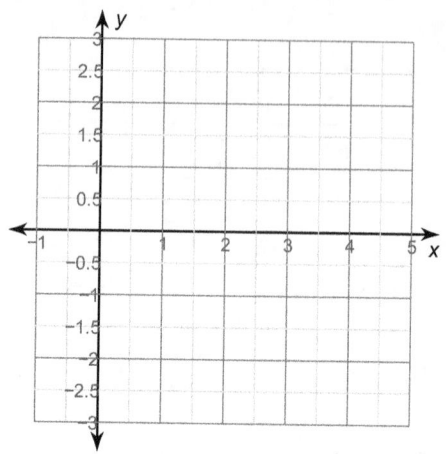

(492) $y = 2(x - 4)^2 + 4$

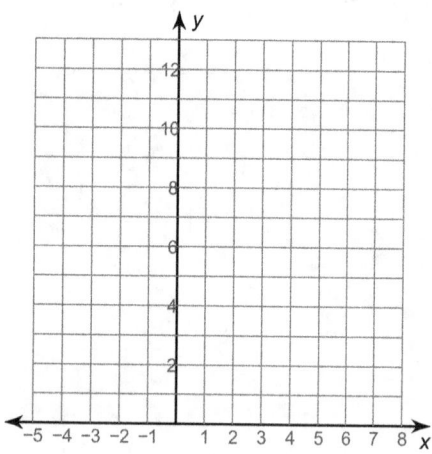

(493) $y = (x - 4)^2 + 1$

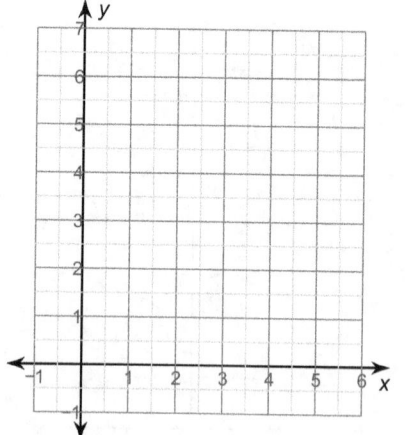

www.math-knots.com

Plot the graph for each of the given functions below.

(494) $y = 2(x-1)^2 + 4$

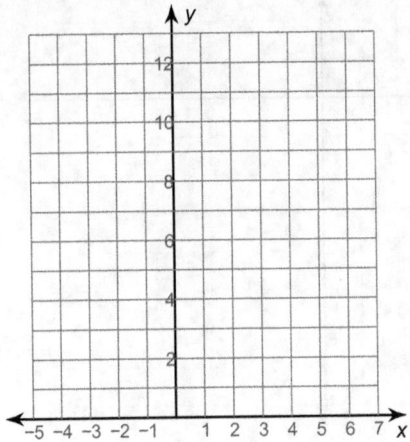

(495) $y = (x+3)^2 + 4$

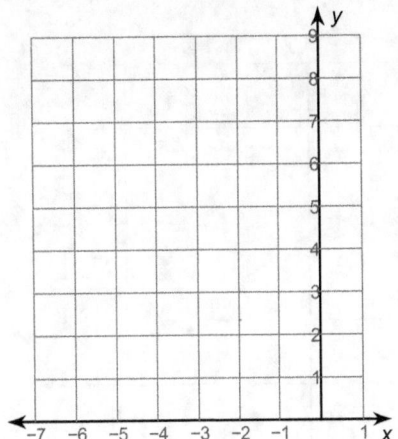

(496) $y = -(x-4)^2 + 4$

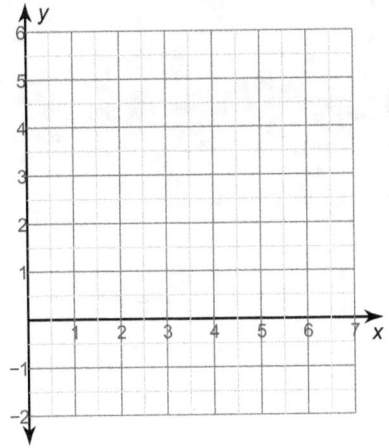

(497) $y = \dfrac{1}{3}(x+3)^2 + 2$

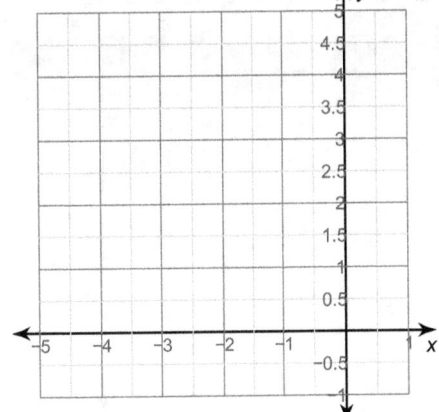

Plot the graph for each of the given functions below.

(498) $y = \dfrac{1}{2}(x-2)^2 + 1$

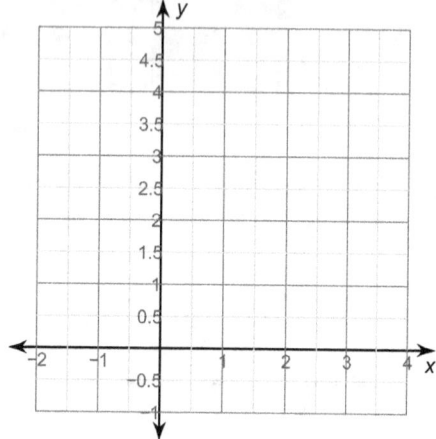

(499) $y = -2(x-2)^2 + 1$

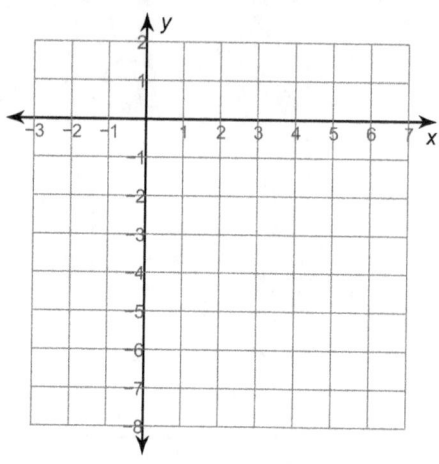

(500) $y = -2(x+4)^2 - 2$

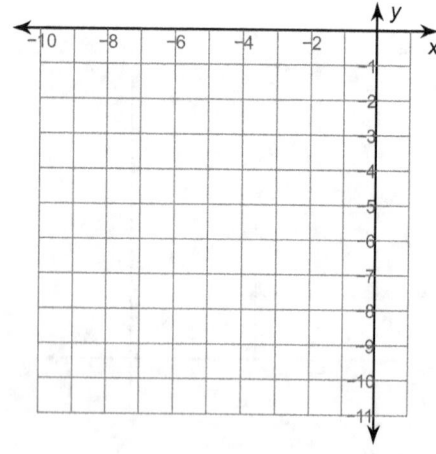

(501) $y = (x-4)^2 - 3$

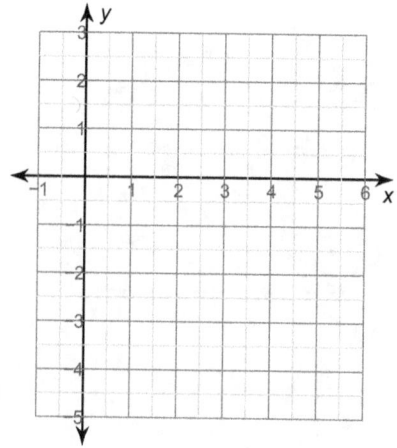

www.math-knots.com

Plot the graph for each of the given functions below.

(502) $y = -2(x + 1)^2 - 1$

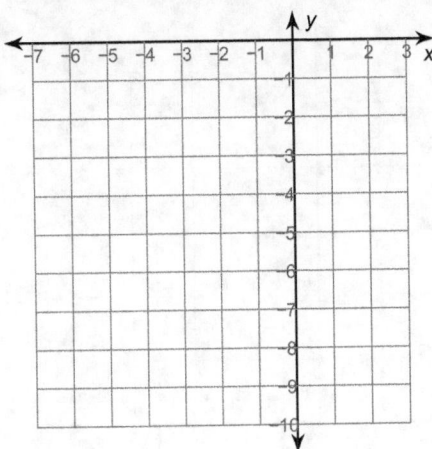

(503) $y = (x - 4)^2 + 4$

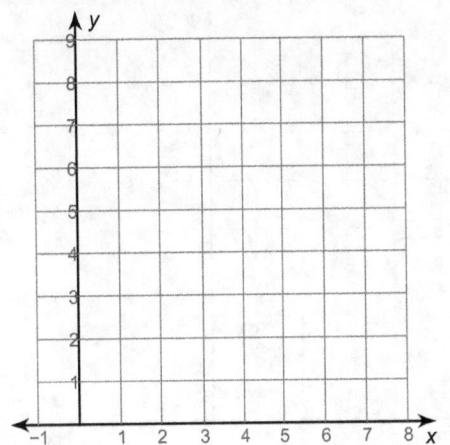

(504) $y = -2(x + 3)^2 - 3$

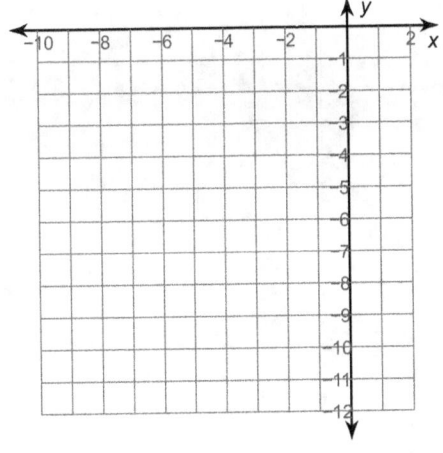

(505) $y = 2(x + 3)^2 + 1$

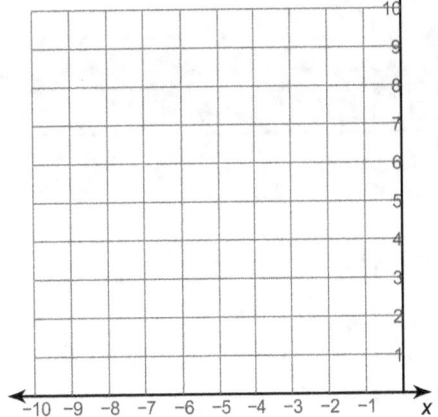

Plot the graph for each of the given functions below.

(506) $y = (x - 1)^2 + 3$

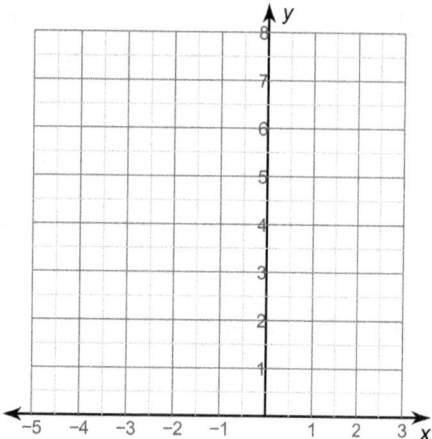

(507) $y = (x - 1)^2 + 1$

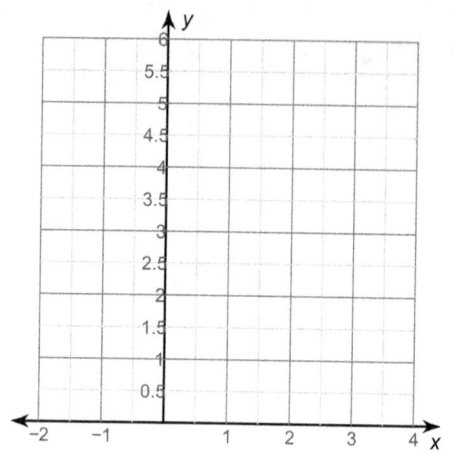

(508) $y = -(x - 2)^2 + 1$

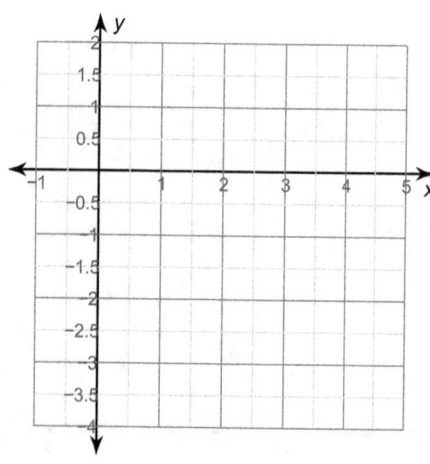

(509) $y = 2(x - 1)^2 + 3$

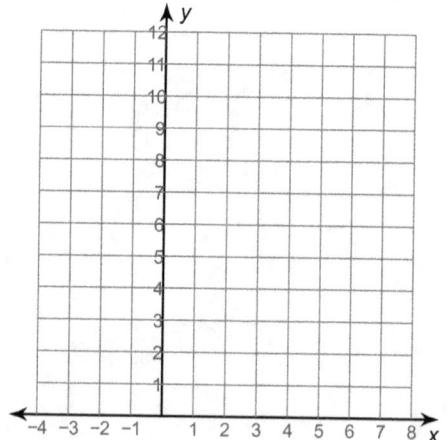

Plot the graph for each of the given functions below.

(510) $y = (x + 3)^2 - 1$

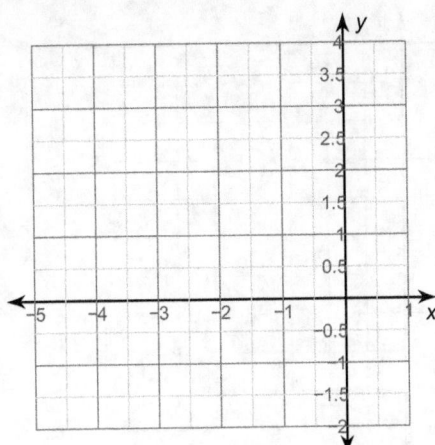

Plot the graph for each of the given inequalities below.

(511) $y \geq -x^2 + 4x - 6$

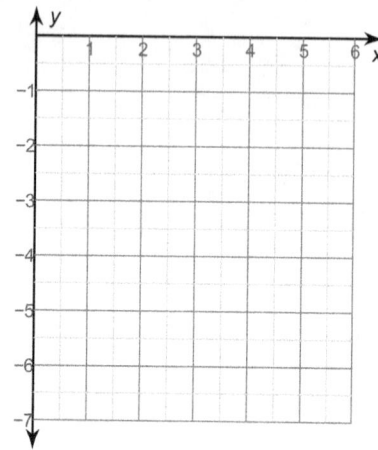

(512) $y > -2x^2 + 16x - 33$

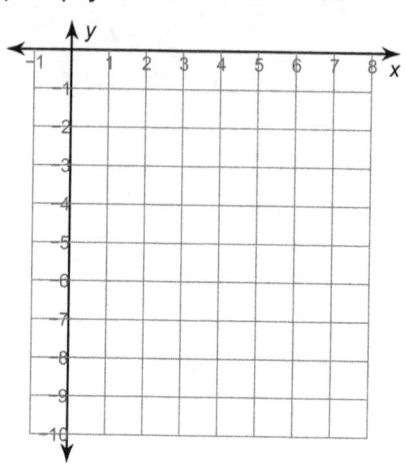

(513) $y \leq 2x^2 + 4x - 1$

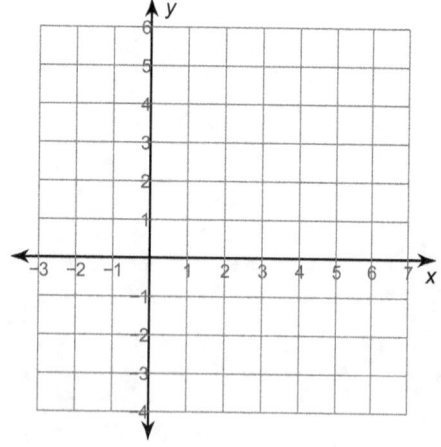

(514) $y \leq -x^2 + 4x - 1$

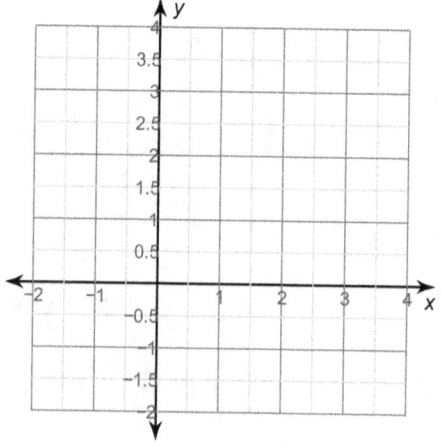

Plot the graph for each of the given inequalities below.

(515) $y \le 2x^2 + 4x$

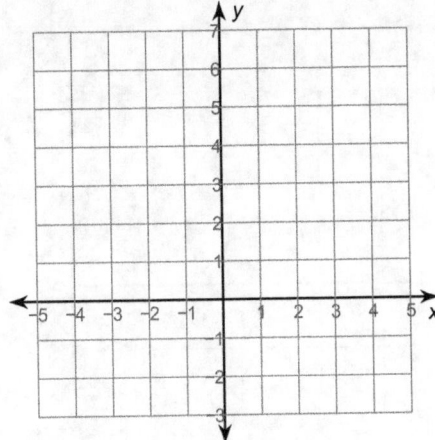

(516) $y \le 2x^2 + 4x + 3$

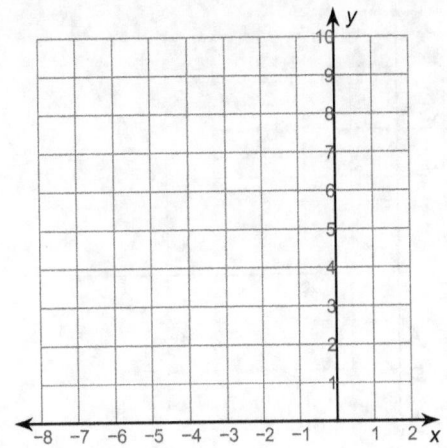

(517) $y > 4x^2 - 16x + 20$

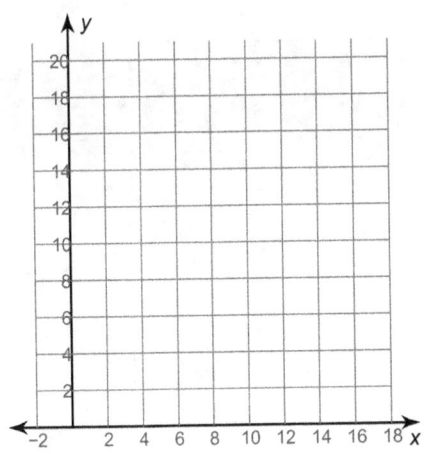

(518) $y < \frac{1}{2}x^2 + 2x - 2$

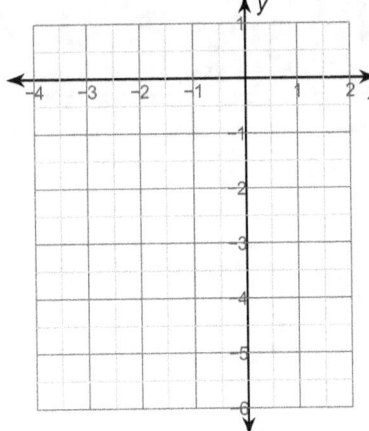

Plot the graph for each of the given inequalities below.

(519) $y > -x^2 + 8x - 13$

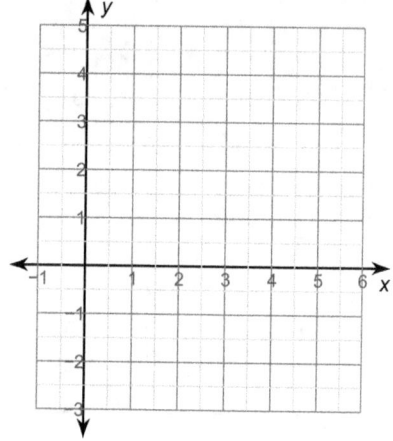

(520) $y > -x^2 - 6x - 6$

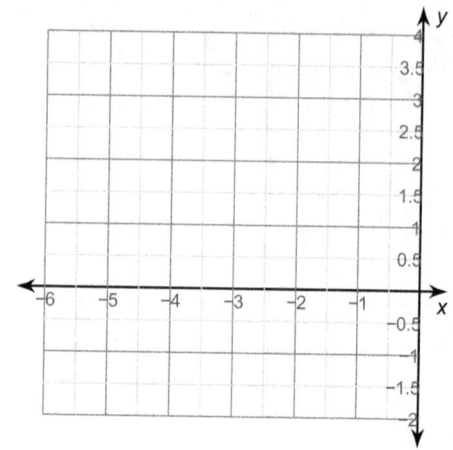

(521) $y > -2x^2 + 8x - 6$

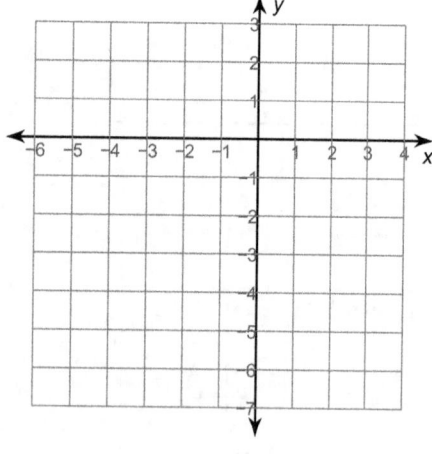

(522) $y < x^2 + 4x + 5$

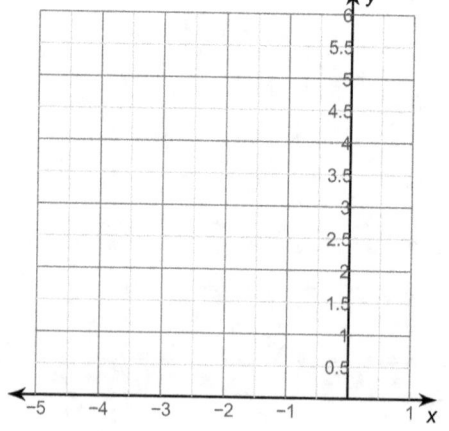

Plot the graph for each of the given inequalities below.

(523) $y < x^2 - 6x + 8$

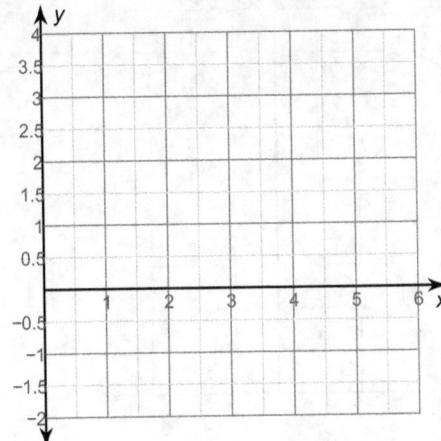

(524) $y > -x^2 - 2x + 2$

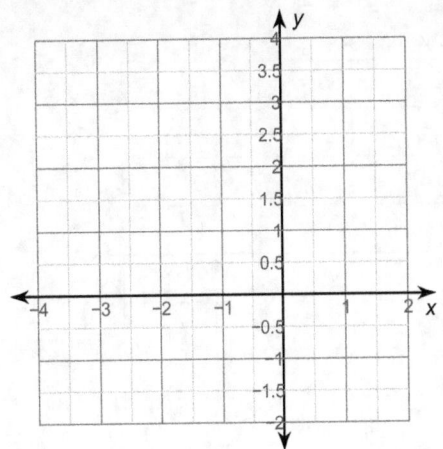

(525) $y < -2x^2 + 8x - 9$

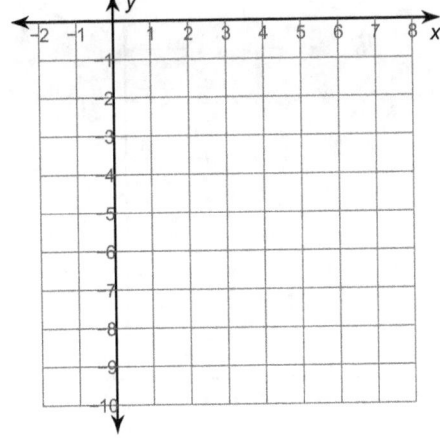

(526) $y > -x^2 + 2x + 3$

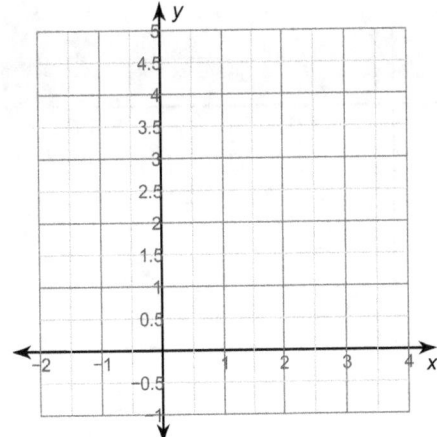

Plot the graph for each of the given inequalities below.

(527) $y < -\dfrac{1}{2}x^2 + 4x - 7$

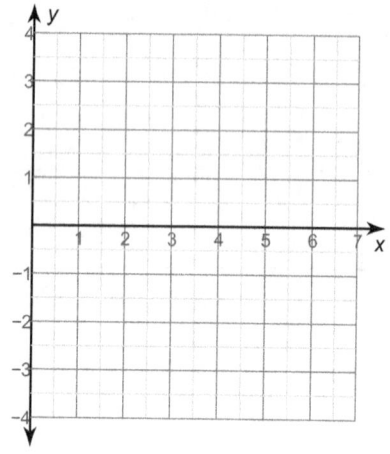

(528) $y < x^2 + 2x + 5$

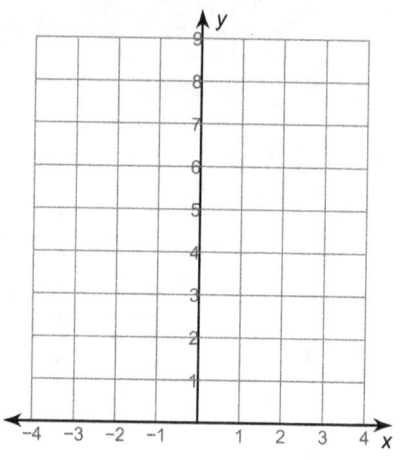

(529) $y < -\dfrac{1}{2}x^2 + 4x - 11$

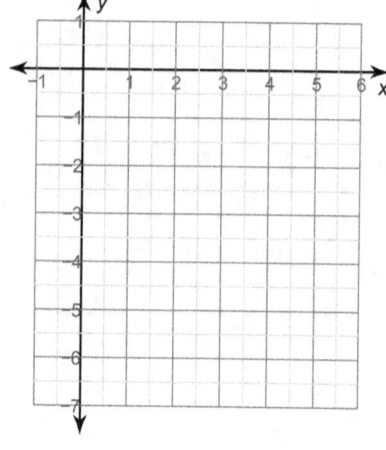

(530) $y \le -\dfrac{1}{2}x^2 - 2x + 2$

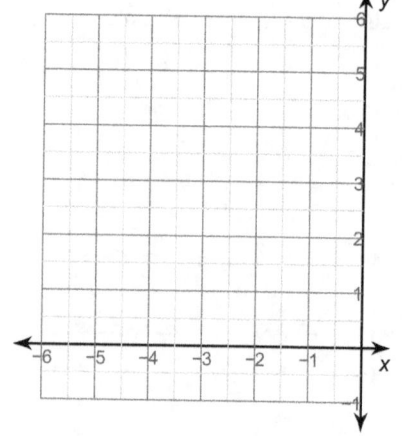

Plot the graph for each of the given inequalities below.

(531) $y > -x^2 - 2x - 4$

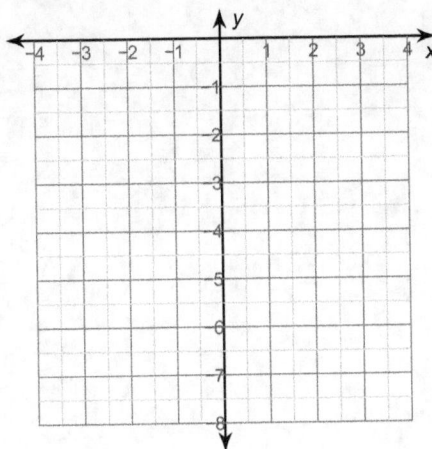

(532) $y < 2x^2 - 4x + 1$

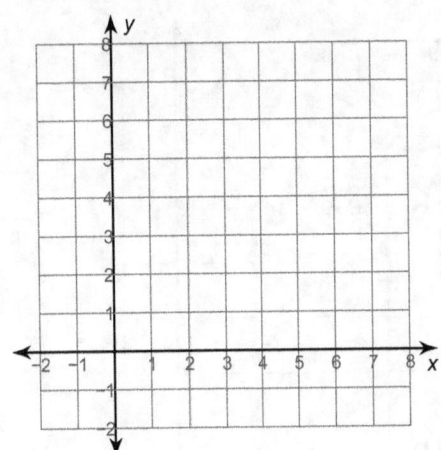

(533) $y \geq 2x^2 + 8x + 9$

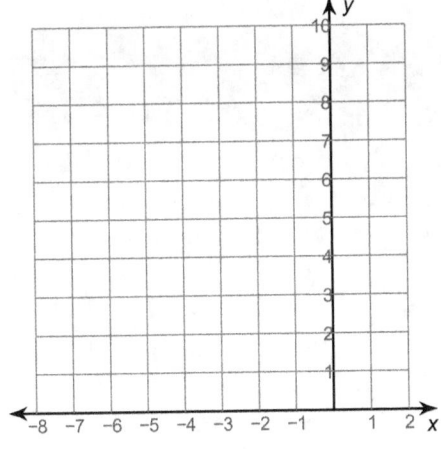

(534) $y \leq 2x^2 + 8x + 6$

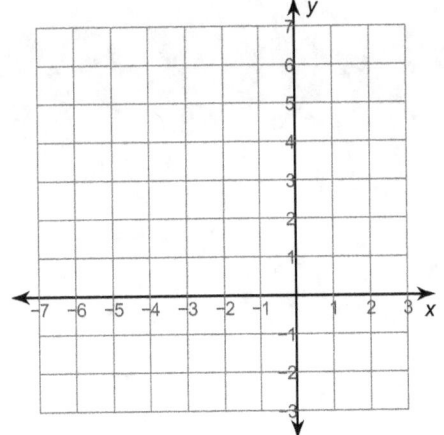

www.math-knots.com

Plot the graph for each of the given inequalities below.

(535) $y > -x^2 - 2x - 3$

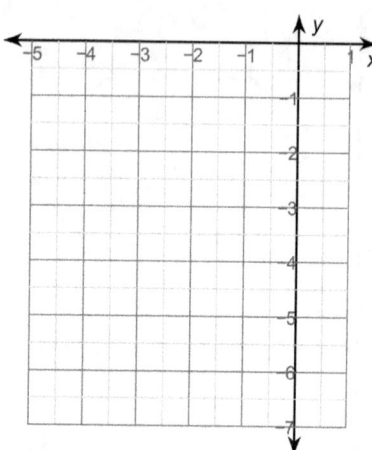

Simplify the below and evaluate the variable by using square roots.

(536) $7v^2 + 5 = 705$

(537) $3 - 9n^2 = -141$

(538) $10v^2 - 5 = 35$

(539) $6n^2 - 7 = -1$

(540) $4p^2 - 3 = 193$

(541) $2x^2 + 9 = 107$

(542) $8x^2 + 5 = 37$

(543) $4a^2 + 4 = 260$

(544) $3x^2 - 10 = 65$

(545) $81n^2 + 4 = 104$

Simplify the below and evaluate the variable by using square roots.

(546) $3b^2 - 7 = 5$

(547) $64v^2 - 10 = 15$

(548) $10n^2 + 10 = 170$

(549) $8x^2 + 1 = 73$

(550) $10n^2 + 6 = 1006$

(551) $36b^2 - 2 = 7$

(552) $-9 - 3x^2 = -309$

(553) $6n^2 + 2 = 56$

(554) $2v^2 - 5 = 93$

(555) $64k^2 - 4 = 45$

Simplify the below and evaluate the variable by using square roots.

(556) $10x^2 + 5 = 1005$

(557) $10n^2 - 9 = 1$

(558) $9m^2 + 3 = 12$

(559) $6r^2 - 3 = 21$

(560) $-1 + 49m^2 = 8$

Find the solution or solutions for the below equations.

(561) $x^2 - 13x + 36 = 0$

(562) $x^2 + 12x + 35 = 0$

(563) $n^2 - 5n - 66 = 0$

(564) $x^2 + 5x + 4 = 0$

(565) $n^2 - 7n = 0$

(566) $x^2 - 49 = 0$

(567) $r^2 + 19r + 84 = 0$

(568) $n^2 - 3n - 18 = 0$

(569) $p^2 - 11p - 12 = 0$

(570) $n^2 - 2n - 35 = 0$

Find the solution or solutions for the below equations.

(571) $m^2 - m - 110 = 0$

(572) $n^2 - 10n + 9 = 0$

(573) $r^2 + 16r + 63 = 0$

(574) $x^2 + 2x - 99 = 0$

(575) $n^2 + 6n = 0$

(576) $r^2 - 12r = 0$

(577) $x^2 + 2x - 120 = 0$

(578) $x^2 - 6x + 8 = 0$

(579) $m^2 - 8m = 0$

(580) $b^2 + 5b = 0$

Find the solution or solutions for the below equations.

(581) $n^2 - 4n = 0$

(582) $x^2 + x - 56 = 0$

(583) $p^2 + 7p - 18 = 0$

(584) $n^2 + 16n + 55 = 0$

(585) $p^2 + 23p + 132 = 0$

(586) $2m^2 + 150m + 1200 = 10m - 2m^2$

(587) $8x^2 - 13x = 504 + 3x$

(588) $v^2 - 20v + 8 = 8$

(589) $34n^2 - 3520 = 18n^2 - 144n$

(590) $20x^2 + 420 = -114x + 14x^2$

Find the solution or solutions for the below equations.

(591) $x^2 + 12x - 120 = 14x$

(592) $p^2 - 20 = -19p$

(593) $5b^2 - 21b + 110 = 4b^2$

(594) $13b^2 + 112b + 12 = 12 - b^2$

(595) $2n^2 - 416 = -16 + n^2$

(596) $-15n^2 + 30n + 200 = -16n^2$

(597) $-2k^2 + 3k - 208 = -3k^2$

(598) $15x^2 + 195x - 1722 = -12$

(599) $r^2 - r = 90$

(600) $11n^2 - 3366 = -11n$

Find the solution or solutions for the below equations.

(601) $7k^2 - 91k + 151 = -3$

(602) $x^2 - 37x = -64 - 17x$

(603) $x^2 - 20x + 92 = 1$

(604) $x^2 - 6x = 135$

(605) $n^2 - 40n + 130 = -17n$

Find the number of solutions for the below equations and its discriminant value.

(606) $-6x^2 + 4x - 2 = 0$

(607) $x^2 - 4x + 4 = 0$

(608) $-5n^2 - 10n - 5 = 0$

(609) $-3b^2 + 3b - 1 = 0$

(610) $-5x^2 - x - 9 = 0$

(611) $-5p^2 + 10p - 5 = 0$

(612) $-10m^2 + 3m + 7 = 0$

(613) $9k^2 + 6k + 1 = 0$

(614) $-x^2 - 4x - 4 = 0$

(615) $3k^2 - 4k + 9 = 0$

Find the number of solutions for the below equations and its discriminant value.

(616) $-5x^2 - 5x + 10 = 0$

(617) $-6x^2 + 2x = 0$

(618) $-2v^2 + 4v - 2 = 0$

(619) $-b^2 + 4b - 7 = 0$

(620) $4b^2 - 10b + 4 = 0$

(621) $4x^2 + 8x + 4 = 0$

(622) $v^2 - 6v + 9 = 0$

(623) $-7v^2 - 8v - 1 = 0$

(624) $10a^2 + 8a - 2 = 0$

(625) $-3x^2 + 6x - 3 = 0$

Find the number of solutions for the below equations and its discriminant value.

(626) $9x^2 - x + 6 = 0$

(627) $-6n^2 + 7n - 1 = 0$

(628) $-10p^2 + 9p - 9 = 0$

(629) $3k^2 + 6k + 3 = 0$

(630) $-4b^2 - 8b - 4 = 0$

107   www.math-knots.com

Find the value of c that completes the square for the given expressions below.

(631) $p^2 + \dfrac{8}{5}p + c$

(632) $p^2 - 20p + c$

(633) $z^2 - \dfrac{4}{13}z + c$

(634) $x^2 - 28x + c$

(635) $x^2 + 32x + c$

(636) $x^2 + \dfrac{11}{17}x + c$

(637) $a^2 + 3a + c$

(638) $m^2 + \dfrac{15}{13}m + c$

(639) $n^2 + \dfrac{127}{12}n + c$

(640) $a^2 - 32a + c$

108   www.math-knots.com

Find the value of c that completes the square for the given expressions below.

(641) $x^2 + 36x + c$

(642) $x^2 + 21x + c$

(643) $z^2 + \dfrac{69}{20}z + c$

(644) $x^2 - 4x + c$

(645) $a^2 + 34a + c$

(646) $x^2 - \dfrac{13}{3}x + c$

(647) $m^2 - 18m + c$

(648) $x^2 - 7x + c$

(649) $x^2 - 42x + c$

(650) $x^2 + 2x + c$

Find the value of c that completes the square for the given expressions below.

(651) $r^2 - 15r + c$

(652) $n^2 + 42n + c$

(653) $x^2 - 12x + c$

(654) $r^2 - 19r + c$

(655) $x^2 - 8x + c$

(656) $x^2 - 17x + c$

(657) $n^2 + \dfrac{35}{18}n + c$

(658) $x^2 - 5x + c$

(659) $x^2 - 38x + c$

(660) $r^2 + \dfrac{13}{16}r + c$

110   www.math-knots.com

Find the value of c that completes the square for the given expressions below.

(661) $a^2 + 28a + c$

(662) $x^2 + 15x + c$

(663) $p^2 - 11p + c$

(664) $y^2 - 34y + c$

(665) $m^2 - \dfrac{5}{3}m + c$

(666) $x^2 - 40x + c$

(667) $z^2 - \dfrac{25}{8}z + c$

(668) $x^2 + 18x + c$

(669) $a^2 + 17a + c$

(670) $x^2 + 11x + c$

Find the value of c that completes the square for the given expressions below.

(671) $a^2 + 26a + c$

(672) $x^2 - 3x + c$

(673) $n^2 + 5n + c$

(674) $m^2 + 12m + c$

(675) $z^2 - 16z + c$

(676) $a^2 - 26a + c$

(677) $a^2 + 14a + c$

(678) $n^2 - 6n + c$

(679) $z^2 - \dfrac{2}{5}z + c$

(680) $r^2 + 7r + c$

www.math-knots.com

Solve the below equations by completing the square.

(681) $x^2 + 2x - 49 = 2$

(682) $a^2 + 16a + 42 = 3$

683) $k^2 - 8k - 10 = 10$

(684) $n^2 - 10n + 22 = 6$

(685) $a^2 + 20a - 24 = -9$

(686) $n^2 + 6n + 15 = 7$

(687) $k^2 + 10k + 22 = 6$

(688) $x^2 + 8x - 73 = -8$

(689) $m^2 + 2m - 12 = 7$

(690) $n^2 - 4n - 91 = -3$

Solve the below equations by completing the square.

(691) $m^2 + 8m - 5 = 4$

(692) $n^2 + 4n - 80 = -7$

(693) $b^2 + 2b - 75 = 5$

(694) $r^2 - 18r + 70 = 5$

(695) $n^2 - 14n + 47 = 2$

(696) $x^2 + 2x = 8$

(697) $m^2 + 16m + 37 = 9$

(698) $r^2 - 2r - 21 = -6$

(699) $r^2 + 14r + 48 = 7$

(700) $x^2 - 4x - 16 = -4$

Solve the below equations by completing the square.

(701) $p^2 - 16p - 90 = -10$

(702) $n^2 - 10n - 34 = 5$

(703) $x^2 - 10x - 21 = 2$

(704) $n^2 - 20n + 23 = 4$

(705) $x^2 - 6x - 87 = 4$

Simplify the below to its lowest possible terms.

(706) $\sqrt{392\,p^2}$

(707) $\sqrt{54\,x^3}$

(708) $\sqrt{72\,p^3}$

(709) $\sqrt{50\,v^4}$

(710) $\sqrt{150\,n}$

(711) $\sqrt{128\,x^3}$

(712) $\sqrt{192\,m^2}$

(713) $\sqrt{72\,b^4}$

(714) $\sqrt{80\,b^3}$

(715) $\sqrt{8\,n^2}$

www.math-knots.com

Simplify the below to its lowest possible terms.

(716) $\sqrt{72x}$

(717) $\sqrt{128x}$

(718) $\sqrt{196m^3}$

(719) $\sqrt{343v^2}$

(720) $\sqrt{98a^2}$

(721) $\sqrt{112p^4}$

(722) $\sqrt{32x^2}$

(723) $\sqrt{24x}$

(724) $\sqrt{448x^2}$

(725) $\sqrt{20p^4}$

Simplify the below to its lowest possible terms.

(726) $\sqrt{250\,x^4 y^4}$

(727) $\sqrt{1008\,x^2 y^2}$

(728) $\sqrt{144\,m^5 n^7}$

(729) $\sqrt{588\,u^6 v^6}$

(730) $\sqrt{1152\,u^7 v^4}$

(731) $\sqrt{192\,u^7 v}$

(732) $\sqrt{192\,ab^3}$

(733) $\sqrt{216\,x^6 y^2}$

(734) $\sqrt{64\,x^5 y^4}$

(735) $\sqrt{1500\,u^6 v^2}$

Simplify the below to its lowest possible terms.

(736) $\sqrt{96\,a^3 b^6}$

(737) $\sqrt{48\,u^6 v^7}$

(738) $\sqrt{1372\,xy^7}$

(739) $\sqrt{150\,x^5 y^7}$

(740) $\sqrt{1690\,u^2 v}$

(741) $\sqrt{54\,xy}$

(742) $\sqrt{24\,m^2 n^7}$

(743) $\sqrt{1372\,x^3 y^2}$

(744) $\sqrt{243\,m^2 n^6}$

(745) $\sqrt{784\,x^3 y^4}$

119
www.math-knots.com

Simplify the below to its lowest possible terms.

(746) $\sqrt{72\,p^4q^4r}$

(747) $\sqrt{80\,xy^2z^4}$

(748) $\sqrt{98\,hjk^3}$

(749) $\sqrt{144\,m^4p^4q^2}$

(750) $\sqrt{252\,a^3b^3c}$

(751) $\sqrt{20\,m^3p^2q^4}$

(752) $\sqrt{50\,m^4p^3q}$

(753) $\sqrt{144\,x^2y^2z^3}$

(754) $\sqrt{196\,x^2y^4z^3}$

(755) $\sqrt{16\,x^4y^4z}$

Simplify the below to its lowest possible terms.

(756) $\sqrt{125\, p^3 q^2 r^4}$

(757) $\sqrt{216\, x^2 y^3 z}$

(758) $\sqrt{27\, m^3 n p}$

(759) $\sqrt{252\, x^4 y z^2}$

(760) $\sqrt{45\, p q^2 r}$

(761) $\sqrt{100\, x^3 y^2 z}$

(762) $\sqrt{196\, x y^2 z^2}$

(763) $\sqrt{96\, m p^2 q^3}$

(764) $\sqrt{24\, m^2 n p^4}$

(765) $\sqrt{256\, x^2 y^4 z^2}$

www.math-knots.com

Simplify the below to its lowest possible terms.

(766) $2\sqrt{27} - 3\sqrt{12} - 2\sqrt{27}$

(767) $2\sqrt{20} - 2\sqrt{20} + 2\sqrt{5}$

(768) $3\sqrt{24} + 2\sqrt{6} - \sqrt{5}$

(769) $-2\sqrt{27} - \sqrt{6} - 3\sqrt{27}$

(770) $-3\sqrt{20} + 3\sqrt{45} - \sqrt{24}$

(771) $-3\sqrt{24} + 2\sqrt{24} - \sqrt{54}$

(772) $3\sqrt{2} - 2\sqrt{3} + 3\sqrt{3}$

(773) $2\sqrt{8} - \sqrt{3} - \sqrt{2}$

(774) $-\sqrt{5} - 2\sqrt{8} + 2\sqrt{45}$

(775) $-2\sqrt{18} - 3\sqrt{8} - 2\sqrt{18}$

122   www.math-knots.com

Simplify the below to its lowest possible terms.

(776) $-3\sqrt{45} - 2\sqrt{6} - 2\sqrt{5}$

(777) $-3\sqrt{12} - 3\sqrt{18} - 2\sqrt{18}$

(778) $-\sqrt{27} + 2\sqrt{20} + 2\sqrt{20}$

(779) $2\sqrt{8} - \sqrt{45} + 2\sqrt{45}$

(780) $-\sqrt{12} - 3\sqrt{27} - \sqrt{6}$

(781) $2\sqrt{45} + 2\sqrt{8} - \sqrt{2}$

(782) $3\sqrt{20} + 3\sqrt{5} + 3\sqrt{2}$

(783) $2\sqrt{3} - 2\sqrt{12} + 3\sqrt{8}$

(784) $-3\sqrt{45} + 3\sqrt{5} + 3\sqrt{5}$

(785) $-3\sqrt{45} + 2\sqrt{5} - 3\sqrt{5}$

Simplify the below to its lowest possible terms.

(786) $-\sqrt{8} - 2\sqrt{5} + 2\sqrt{2}$

(787) $-\sqrt{5} - 3\sqrt{5} - 2\sqrt{5}$

(788) $-2\sqrt{3} + 2\sqrt{27} - 2\sqrt{27}$

(789) $3\sqrt{18} - \sqrt{2} - 2\sqrt{12}$

(790) $2\sqrt{6} + 3\sqrt{54} - 2\sqrt{18}$

(791) $\sqrt{15}(\sqrt{2} + \sqrt{3})$

(792) $\sqrt{15}(2 + \sqrt{3})$

(793) $\sqrt{10}(-4\sqrt{6} + \sqrt{5})$

(794) $\sqrt{6}(\sqrt{10} + \sqrt{2})$

(795) $4\sqrt{10}(\sqrt{5} + 4\sqrt{2})$

Simplify the below to its lowest possible terms.

(796) $\sqrt{15}\left(5 + 5\sqrt{6}\right)$

(797) $\sqrt{15}\left(\sqrt{10} + 5\right)$

(798) $2\sqrt{5}\left(\sqrt{10} + \sqrt{6}\right)$

(799) $3\sqrt{3}\left(\sqrt{10} + 5\right)$

(800) $\sqrt{10}\left(3 + \sqrt{2}\right)$

(801) $\sqrt{6}\left(\sqrt{3} + \sqrt{5}\right)$

(802) $2\sqrt{3}\left(\sqrt{2} - \sqrt{3}\right)$

(803) $\sqrt{10}\left(4 + \sqrt{2}\right)$

(804) $\sqrt{15}\left(\sqrt{2} - 4\sqrt{10}\right)$

(805) $\sqrt{10}\left(-2\sqrt{10} + \sqrt{2}\right)$

Simplify the below to its lowest possible terms.

(806) $5\sqrt{5}\left(4+\sqrt{10}\right)$

(807) $\sqrt{15}\left(\sqrt{3}+5\right)$

(808) $\sqrt{2}\left(\sqrt{10}+\sqrt{3}\right)$

(809) $\sqrt{6}\left(-2\sqrt{3}+\sqrt{5}\right)$

(810) $-\sqrt{6}\left(-3\sqrt{2}+3\sqrt{5}\right)$

(811) $-\dfrac{2}{5\sqrt{2}}$

(812) $\dfrac{\sqrt{4}}{\sqrt{5}}$

(813) $\dfrac{\sqrt{3}}{\sqrt{5}}$

(814) $-\dfrac{4}{\sqrt{3}}$

(815) $\dfrac{\sqrt{5}}{2\sqrt{3}}$

Simplify the below to its lowest possible terms.

(816) $\dfrac{4\sqrt{5}}{\sqrt{3}}$

(817) $\dfrac{\sqrt{9}}{3\sqrt{15}}$

(818) $\dfrac{\sqrt{2}}{\sqrt{5}}$

(819) $\dfrac{5}{\sqrt{3}}$

(820) $\dfrac{\sqrt{5}}{\sqrt{2}}$

(821) $\dfrac{\sqrt{3}-\sqrt{2}}{\sqrt{6}}$

(822) $\dfrac{2\sqrt{5}-\sqrt{2}}{2\sqrt{5}}$

(823) $\dfrac{5+\sqrt{3}}{\sqrt{7}}$

(824) $\dfrac{-2+\sqrt{2}}{\sqrt{12}}$

(825) $\dfrac{-2+5\sqrt{5}}{\sqrt{5}}$

Simplify the below to its lowest possible terms.

(826) $\dfrac{2 + 4\sqrt{2}}{\sqrt{12}}$

(827) $\dfrac{3 - \sqrt{3}}{\sqrt{8}}$

(828) $\dfrac{-4 - \sqrt{3}}{3\sqrt{19}}$

(829) $\dfrac{4 + 4\sqrt{5}}{\sqrt{13}}$

(830) $\dfrac{4 + \sqrt{2}}{4\sqrt{14}}$

(831) $\dfrac{4 - \sqrt{2}}{5\sqrt{11}}$

(832) $\dfrac{\sqrt{3} - 4\sqrt{5}}{3\sqrt{5}}$

(833) $\dfrac{3 - 3\sqrt{3}}{\sqrt{20}}$

(834) $\dfrac{\sqrt{5} + 3\sqrt{3}}{\sqrt{13}}$

(835) $\dfrac{-5 + 3\sqrt{3}}{3\sqrt{5}}$

Simplify the below to its lowest possible terms.

(836) $\dfrac{\sqrt{2} + 2\sqrt{5}}{5 - 3\sqrt{3}}$

(837) $\dfrac{-1 - 3\sqrt{5}}{\sqrt{2} + 4\sqrt{5}}$

(838) $\dfrac{5\sqrt{5} + \sqrt{3}}{4 - \sqrt{3}}$

(839) $\dfrac{4 - \sqrt{3}}{3 - 2\sqrt{2}}$

(840) $\dfrac{3 + \sqrt{3}}{2 + \sqrt{3}}$

(841) $\dfrac{4 + \sqrt{2}}{4 + 2\sqrt{2}}$

(842) $\dfrac{-2 + 4\sqrt{3}}{5 - 5\sqrt{3}}$

(843) $\dfrac{-4 - 2\sqrt{2}}{3 - 4\sqrt{2}}$

(844) $\dfrac{3 - 2\sqrt{5}}{5 - 5\sqrt{2}}$

(845) $\dfrac{-3 + \sqrt{5}}{-1 + \sqrt{2}}$

Simplify the below to its lowest possible terms.

(846) $\dfrac{2 + \sqrt{3}}{5\sqrt{5} + \sqrt{3}}$

(847) $\dfrac{4 - 2\sqrt{2}}{5\sqrt{3} + \sqrt{2}}$

(848) $\dfrac{2 + \sqrt{2}}{5 - 5\sqrt{5}}$

(849) $\dfrac{-4 - 3\sqrt{2}}{-3 - \sqrt{3}}$

(850) $\dfrac{-5 - 5\sqrt{2}}{3\sqrt{3} + 5}$

(851) $\dfrac{-2 - \sqrt{5}}{3 + \sqrt{3}}$

(852) $\dfrac{4 + 4\sqrt{3}}{-4 - 3\sqrt{3}}$

(853) $\dfrac{5 + 4\sqrt{5}}{3 - \sqrt{2}}$

(854) $\dfrac{5 + 5\sqrt{2}}{-2 + \sqrt{2}}$

(855) $\dfrac{5 + \sqrt{2}}{5 - 5\sqrt{5}}$

Simplify the below to its lowest possible terms.

(856) $\dfrac{3\sqrt{2} - 5}{4 + 5\sqrt{2}}$

(857) $\dfrac{4 - \sqrt{5}}{5 - \sqrt{2}}$

(858) $\dfrac{5 - 5\sqrt{5}}{5 + \sqrt{3}}$

(859) $\dfrac{-4 + \sqrt{3}}{3 - \sqrt{2}}$

(860) $\dfrac{-5 + 3\sqrt{5}}{\sqrt{5} + \sqrt{2}}$

(861) $\dfrac{3 - 3\sqrt{5}}{4 + 5\sqrt{5}}$

(862) $\dfrac{5 - 4\sqrt{3}}{-1 + 2\sqrt{5}}$

(863) $\dfrac{-4 + 4\sqrt{2}}{2 - 2\sqrt{2}}$

(864) $\dfrac{2 - \sqrt{2}}{2 - 3\sqrt{2}}$

(865) $\dfrac{-2 + \sqrt{3}}{2 - 5\sqrt{2}}$

Evaluate the below equation. Check for extraneous solutions.

(866) $2\sqrt{n+10} = 4$

(867) $4 = \sqrt{\dfrac{b}{10}}$

(868) $8 + \sqrt{37n-1} = 14$

(869) $-1 + \sqrt{\dfrac{n}{5}} = -1$

(870) $-7\sqrt{11-2b} = -21$

(871) $14 = 10 + \sqrt{6v-2}$

(872) $15 = 3\sqrt{\dfrac{r}{3}}$

(873) $6 = \sqrt{18-2x} + 4$

(874) $-9\sqrt{20m+4} = -72$

(875) $\sqrt{\dfrac{k}{6}} = 0$

Evaluate the below equation. Check for extraneous solutions.

(876) $0 = -2\sqrt{4 - 2r}$

(877) $\sqrt{1 - 15x} + 3 = 7$

(878) $5\sqrt{x + 7} = 15$

(879) $9 = \sqrt{9x}$

(880) $7 = \sqrt{n - 10}$

(881) $\sqrt{7 - 2v} = \sqrt{5v}$

(882) $\sqrt{3x} = \sqrt{2x + 3}$

(883) $\sqrt{15 - x} = \sqrt{2x - 6}$

(884) $\sqrt{6n - 1} = \sqrt{5n}$

(885) $\sqrt{\dfrac{n}{9}} = \sqrt{20 - n}$

Evaluate the below equation. Check for extraneous solutions.

(886) $\sqrt{\dfrac{n}{3}} = \sqrt{12 - n}$

(887) $\sqrt{3a - 3} = \sqrt{2a}$

(888) $\sqrt{5n} = \sqrt{2n + 6}$

(889) $\sqrt{12 - 2a} = \sqrt{2a}$

(890) $\sqrt{3n - 42} = \sqrt{\dfrac{n}{5}}$

(891) $x = \sqrt{30 - x}$

(892) $v = \sqrt{-36 + 13v}$

(893) $p = \sqrt{-5 + 6p}$

(894) $\sqrt{-6 + 7x} = x$

(895) $b = \sqrt{-16 + 10b}$

134   www.math-knots.com

ADVANCED ALGEBRA 1

Volume 2

Evaluate the below equation. Check for extraneous solutions.

(896) $\sqrt{8b} = b$

(897) $a = \sqrt{56 - a}$

(898) $r = \sqrt{-36 + 13r}$

(899) $\sqrt{72 - v} = v$

(900) $n = \sqrt{2 - n}$

(901) $r = 5 + \sqrt{5r - 31}$

(902) $\sqrt{58 - 6x} = x - 7$

(903) $\sqrt{4n + 8} = n + 3$

(904) $\sqrt{2a - 14} = a - 7$

(905) $n - 7 = \sqrt{41 - 5n}$

www.math-knots.com

Evaluate the below equation. Check for extraneous solutions.

(906) $2 = -x + \sqrt{4x + 5}$

(907) $\sqrt{4x - 3} = x - 2$

(908) $k + 3 = \sqrt{5k + 51}$

(909) $\sqrt{4b - 8} = b - 1$

(910) $m - 1 = \sqrt{7 - 7m}$

Simplify the below expressions.

(911) $\dfrac{8n+12}{4} \cdot \dfrac{6n^2-12n}{18n+27}$

(912) $\dfrac{3v^3+3v^2}{3v^3-3v^2} \cdot \dfrac{8-7v-v^2}{v^2+9v+8}$

(913) $\dfrac{b^2-b-72}{b-9} \cdot \dfrac{7b+56}{7b+70}$

(914) $\dfrac{n^2-15n+50}{10-n} \cdot \dfrac{6n^2-24n}{n^2-9n+20}$

(915) $\dfrac{n^2-3n-70}{10-n} \cdot \dfrac{n^2+4n-32}{2n^3+14n^2}$

(916) $\dfrac{2n+20}{4} \cdot \dfrac{4n-8}{n^2+13n+30}$

(917) $\dfrac{8n-32}{n+4} \cdot \dfrac{n^2+9n+20}{8n+40}$

(918) $\dfrac{x^2-2x-80}{4x^2+32x} \cdot \dfrac{4x^2+12x}{7x-70}$

(919) $\dfrac{5p-15p^2}{p+5} \cdot \dfrac{10p+50}{24p-8}$

(920) $\dfrac{16k^3-16k^2}{k^2+19k+90} \cdot \dfrac{5k^3+45k^2}{10k^3-10k^2}$

Simplify the below expressions.

(921) $\dfrac{v^2 - 3v - 4}{4 + 3v - v^2} \cdot \dfrac{v^2 + 8v - 20}{5v + 50}$

(922) $\dfrac{4p + 32}{p^2 - 64} \cdot \dfrac{8p - 64}{4}$

(923) $\dfrac{9b + 9}{27 - 45b} \cdot \dfrac{20b - 12}{b^2 + 3b + 2}$

(924) $\dfrac{p^2 + 9p + 20}{21p^3 - 56p^2} \cdot \dfrac{21p^3 - 56p^2}{6p}$

(925) $\dfrac{r + 8}{r^2 + 6r - 16} \cdot \dfrac{r^2 + 7r - 18}{r^2 + 10r + 25}$

(926) $\dfrac{r^2 + 6r - 27}{r^2 + 5r - 24} \cdot \dfrac{r^2 + 9r + 8}{2r + 18}$

(927) $\dfrac{56 - b - b^2}{b^2 + 16b + 64} \cdot \dfrac{3b^3 + 24b^2}{7b^2 - 49b}$

(928) $\dfrac{n + 5}{n^2 + 4n - 5} \cdot \dfrac{21n^3 + 24n^2}{56n^2 + 64n}$

(929) $\dfrac{7r^3 - 70r^2}{7r^3 - 42r^2} \cdot \dfrac{r^2 - r - 30}{r - 10}$

(930) $\dfrac{k^2 - 4k - 45}{35k^2 - 25k} \cdot \dfrac{35k^2 - 25k}{10k^3 + 50k^2}$

138   www.math-knots.com

Simplify the below expressions.

(931) $\dfrac{4}{m-3} + \dfrac{6m}{m-4}$

(932) $\dfrac{5}{r-6} + \dfrac{6}{4r^3}$

(933) $\dfrac{3r}{2} + \dfrac{r+5}{3r(r+6)}$

(934) $\dfrac{3}{6k^2} + \dfrac{2k}{k-1}$

(935) $\dfrac{3b}{3} + \dfrac{3b}{3(b-2)}$

(936) $\dfrac{5r}{3r} + \dfrac{6r+1}{3(3r-4)}$

(937) $\dfrac{6r}{r+3} + \dfrac{4}{r+4}$

(938) $\dfrac{4m}{3} + \dfrac{m-1}{2(m+6)}$

(939) $\dfrac{6}{2} + \dfrac{6a}{(a+6)(3a+2)}$

(940) $\dfrac{n-3}{3n+1} + \dfrac{2}{5}$

Simplify the below expressions.

(941) $\dfrac{2b}{3} + \dfrac{5}{3b(b-2)}$

(942) $\dfrac{3b}{3(b+1)} + \dfrac{6b}{b-1}$

(943) $\dfrac{5r}{2r(r-6)} + \dfrac{5}{2r}$

(944) $\dfrac{5}{6} + \dfrac{v-6}{v-2}$

(945) $\dfrac{a-6}{6(a-3)} + \dfrac{4a}{3a}$

(946) $\dfrac{4}{k-4} + \dfrac{2}{k-3}$

(947) $\dfrac{5}{x+6} + \dfrac{4}{x-2}$

(948) $\dfrac{4}{r-1} + \dfrac{2}{5r}$

(949) $\dfrac{2}{3r(r+2)} + \dfrac{2}{2r}$

(950) $\dfrac{6}{k+6} + \dfrac{6k}{k+3}$

Simplify the below expressions.

(951) $\dfrac{6x-4}{5x+10} - \dfrac{7}{x-5}$

(952) $\dfrac{6}{4} - \dfrac{3n-9}{3n+18}$

(953) $\dfrac{x-7}{x+5} - \dfrac{3}{6}$

(954) $\dfrac{8}{a-10} - \dfrac{4a}{3a^2}$

(955) $10b - \dfrac{3b}{63b^3 - 54b^2}$

(956) $\dfrac{7}{2} - \dfrac{6}{6n+12}$

(957) $\dfrac{6}{x-7} - \dfrac{3}{8x}$

(958) $\dfrac{3}{k-5} - \dfrac{3}{k+2}$

(959) $\dfrac{8}{p-6} - \dfrac{4}{p-10}$

(960) $\dfrac{7}{3} - \dfrac{r-8}{2r^2 + 4r}$

141   www.math-knots.com

Simplify the below expressions.

(961) $\dfrac{x+6}{14x^2+20x} - \dfrac{5}{2}$

(962) $\dfrac{7}{x^2+x-72} - 6x$

(963) $\dfrac{8}{2} - \dfrac{3}{2a^2-14a}$

(964) $\dfrac{3}{n+9} - \dfrac{9n}{n-5}$

(965) $\dfrac{5b}{3} - \dfrac{b+6}{28b-32}$

(966) $\dfrac{2}{k-2} - \dfrac{4}{k-5}$

(967) $\dfrac{x-10}{x-2} - \dfrac{8}{7x}$

(968) $\dfrac{8}{2x-4} - \dfrac{9x}{9}$

(969) $\dfrac{2}{5n} - \dfrac{n-10}{14n-4}$

(970) $\dfrac{7k}{5k-15} - \dfrac{2k}{4k}$

Simplify the below expressions. Check for the extraneous solutions.

(971) $\dfrac{2}{n^2} + \dfrac{2}{n} = \dfrac{1}{n}$

(972) $\dfrac{x+3}{6x} = \dfrac{1}{6} - \dfrac{1}{2}$

(973) $\dfrac{1}{2x} = \dfrac{1}{2x^2} + \dfrac{1}{x}$

(974) $\dfrac{6}{x^2} = \dfrac{1}{x^2} - \dfrac{2}{x}$

(975) $\dfrac{1}{2k} = \dfrac{1}{4k} + \dfrac{4k-6}{k^2}$

(976) $1 - \dfrac{1}{3} = \dfrac{4a-16}{a}$

(977) $\dfrac{1}{6} = \dfrac{m-2}{6m} + \dfrac{5m-30}{6m}$

(978) $\dfrac{2}{n} = \dfrac{n-2}{n} - \dfrac{1}{n}$

(979) $\dfrac{2}{a^2} + \dfrac{1}{a} = \dfrac{1}{a^2}$

(980) $\dfrac{1}{2m} + \dfrac{2m-12}{m} = \dfrac{1}{2}$

Simplify the below expressions. Check for the extraneous solutions.

(981) $\dfrac{k-1}{3k^2} = \dfrac{k+6}{3k^2} - \dfrac{1}{3k}$

(982) $\dfrac{1}{3k} + \dfrac{1}{3k^2} = \dfrac{1}{k^2}$

(983) $1 - \dfrac{1}{m} = \dfrac{1}{4m}$

(984) $\dfrac{3}{x^2} + \dfrac{x-4}{6x^2} = \dfrac{1}{3x^2}$

(985) $\dfrac{1}{n} = \dfrac{5}{n} + \dfrac{1}{n^2}$

(986) $\dfrac{1}{p} = \dfrac{2p-8}{p^2} + \dfrac{1}{2p}$

(987) $\dfrac{v+6}{6v^2} = \dfrac{5v+1}{2v^2} + \dfrac{1}{6v}$

(988) $\dfrac{1}{x} = \dfrac{3}{2x} + \dfrac{x+2}{4x^2}$

(989) $\dfrac{1}{x} - \dfrac{3x+1}{3x^2} = \dfrac{x+2}{3x^2}$

(990) $\dfrac{1}{5k} = \dfrac{1}{k^2} + \dfrac{1}{k}$

www.math-knots.com

Find the value of each trigonometric ratio as given.

(991) sin X

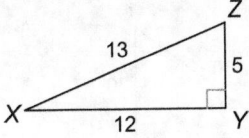

(992) sin C

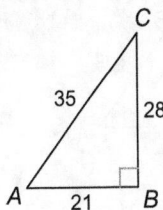

(993) sin A

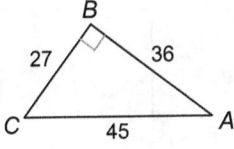

(994) sin A

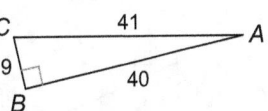

(995) sin A

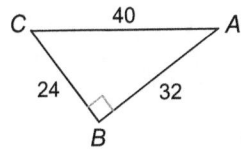

(996) sin Z

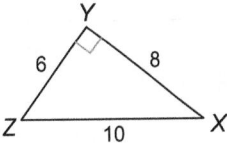

Find the value of each trigonometric ratio as given.

(997) $\sin C$

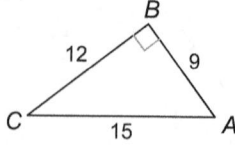

(998) $\sin C$

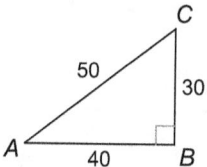

(999) $\sin X$

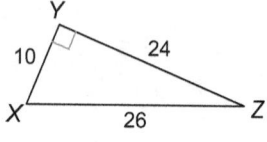

(1000) $\sin X$

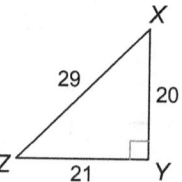

(1001) $\cos C$

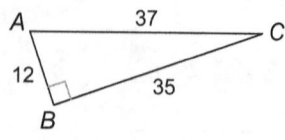

(1002) $\cos C$

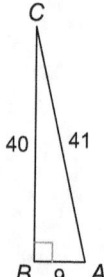

www.math-knots.com

Find the value of each trigonometric ratio as given.

(1003) cosX

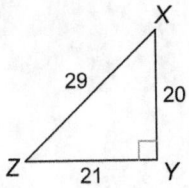

(1004) cosC

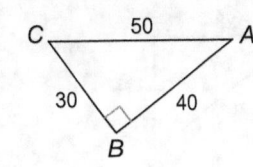

(1005) cosX

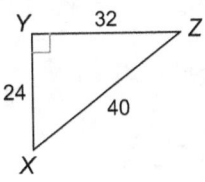

(1006) cosZ

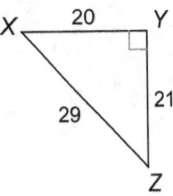

(1007) cosC

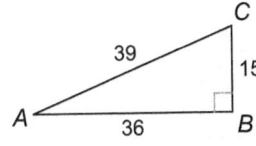

(1008) cosC

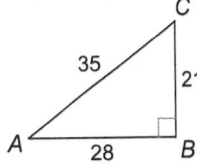

www.math-knots.com

Find the value of each trigonometric ratio as given.

(1009) cos*C*

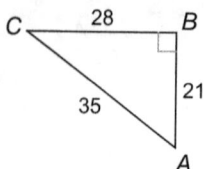

(1010) cos*Z*

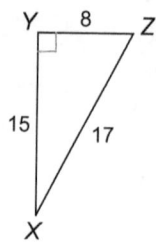

(1011) tan*A*

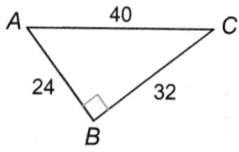

(1012) tan*X*

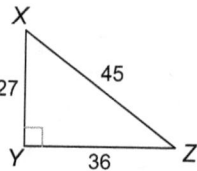

(1013) tan*C*

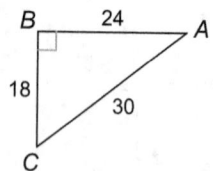

(1014) tan*Z*

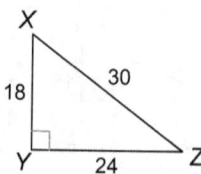

Find the value of each trigonometric ratio as given.

(1015)   tan A

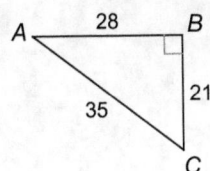

(1016)   tan A

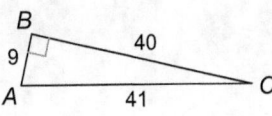

(1017)   tan C

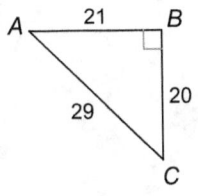

(1018)   tan Z

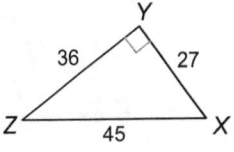

(1019)   tan C

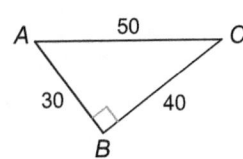

(1020)   tan A

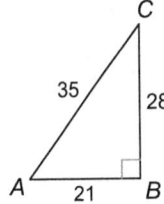

www.math-knots.com

www.math-knots.com

(1)   quadratic trinomial

(2)   sixth degree trinomial

(3)   constant monomial

(4)   eighth degree polynomial
with six terms

(5)   eighth degree polynomial
with four terms

(6)   constant monomial

(7)   linear monomial

(8)   quintic binomial

(9)   seventh degree binomial

(10)   constant monomial

(11)   constant monomial

(12)   cubic binomial

(13)   constant monomial

(14)   quadratic monomial

(15)   seventh degree monomial

(16)   constant monomial

(17)   linear binomial

(18)   quartic polynomial with
four terms

151   www.math-knots.com

(19)   seventh degree monomial

(20)   cubic monomial

(21)   sixth degree monomial

(22)   quadratic binomial

(23)   linear monomial

(24)   cubic monomial

(25)   sixth degree polynomial
       with six terms

(26) $-6x^4 + x^2 + 7x$

(27) $-4n^4 + 5n^2 - 2$

(28) $5n^3$

(29) $x - 3$

(30) $-5n^2$

(31) $-5x^4 + 10x$

(32) $3n^4 - 4n^3 - 2n^2$

(33) $3m^4 + 9m^2$

(34) $b^4 + b^3$

(35) $-10x^4 - 7$

(36) $-15x^4 + 6x^2$

(37) $7x - 2$

(38) $8x^3 - 4$

(39) $v^2 + 6$

(40) $7x - 4$

(41) $9k^4 - 9k^2$

(42) $11p^3 - 1$

(43) $-4p - 5$

(44) $-5n^3 + 8n^2$

(45) $9x^4 + x$

(46) $-11x^2 + 1$

(47) $8x + 5$

(48) $6b$

(49) $5x^2 + 4$

(50) $v^4 + 11v$

(51) $-3b^4 + 5b^3 - b^2 + 3$

(52) $7a^2 - 4a + 8$

(53) $-7n^3 - 7n^2 + 15n + 4$

(54) $2n^3 - 5n^2 + 4$

www.math-knots.com

(55) $-5x^4 - 4x^3 + 3x + 8$

(56) $-x^4 - x^3 + 2x^2$

(57) $-3p^4 + p^3 - 9$

(58) $-5x^2 + 5x + 14$

(59) $13x^2 + 11$

(60) $6p^3 + 16p^2 + 3p - 10$

(61) $12b^3 - 7b^2 - 3b - 7$

(62) $6a^2 + 9$

(63) $8b^3 - 2b^2 + 6b$

(64) $7x^4 + 4x^3 + 5$

(65) $7b^4 + 7b^2 + 13$

(66) $-4b^4 - 3b^3 - 2b^2 + 6b$

(67) $9n^3 - 7n^2 - 6$

(68) $n^3 + 8n^2$

(69) $-14x^4 - 7x^2 - 5x - 7$

(70) $-4k^4 - 3k^2 + 4k$

(71) $8b^3 + 7b + 1$

(72) $-2b^3 + 4b + 10$

(73) $10m^4 - 2m - 10$

(74) $-4v^4 + 5v^3 + 3$

(75) $2x^2 + 5x$

(76) $-17b^3 + 6b^2 + 6b$

(77) $-4n^4 + 6n^3 + 10n^2 + 5n + 6$

(78) $-k^4 - 6k^3 + 3k^2 + 2k + 2$

(79) $-20a^3 + 11a^2 - 9a + 7$

(80) $4b^4 - 6b^3 + 2b^2 - 2b$

(81) $7x^3 - 10x^2 + 12$

(82) $3m^4 - 14m^3 - 12m - 13$

(83) $-9k^4 + 3k^3 + 10k^2 + 5k$

(84) $7n^4 - 2n^3 + 3n^2 + 3n - 2$

(85) $4b^4 + b^2 + b$

(86) $-3m^3 + m - 3$

(87) $9x^4 - 6x^3 + 7x^2 - 1$

(88) $5b^4 - 2b^3 - 10b^2 - 6b + 4$

(89) $-2x^3 + 3x^2 + 8x + 1$

(90) $-5r^4 - 5r^3 + 10r + 5$

(91) $3n^4 + 5n^3 - 2n^2 + 12n + 4$

(92) $-7n^4 - 12n^3 + 3n^2 + 2n + 11$

(93) $10n^4 + 5n^3 - 7n^2 + 7n$

(94) $-2x^4 - 9x^3 + 11x - 6$

(95) $x^4 + 2x^3 - 2x^2 - 2x - 2$

(96) $9a^4 - 6a^3 - a^2 - 12a$

(97) $-7x^4 - 5x^3 - 7x^2 + x + 17$

(98) $-13n^4 - 4n^2 - 13n + 8$

(99) $-9p^3 - 5p^2 + 14p + 11$

(100) $-10n^2 + 2n + 8$

(101) $3x^4 - 10x^3 - 2x + 5$

(102) $-13r^4 + 11r^3 + r + 8$

(103) $-4k^4 - 6k^3 - 14k^2 + 9k + 4$

(104) $-3n^4 + 10n^3 - 13n^2 + 2n + 7$

(105) $-9x^4 - 6x^3 - 14x - 1$

(106) $4 + \dfrac{1}{5n} + \dfrac{2}{5n^2}$

(107) $\dfrac{k^2}{3} + \dfrac{k}{3} + \dfrac{1}{2}$

(108) $\dfrac{x}{4} + \dfrac{3}{8} + \dfrac{3}{8x}$

www.math-knots.com

(109) $\dfrac{2r^2}{5} + 5r + \dfrac{1}{2}$

(110) $\dfrac{k}{3} + \dfrac{1}{2} + \dfrac{3}{k}$

(111) $\dfrac{2k}{3} + \dfrac{1}{2} + \dfrac{5}{k}$

(112) $\dfrac{k^2}{2} + \dfrac{k}{2} + 1$

(113) $\dfrac{x^2}{2} + \dfrac{x}{3} + \dfrac{5}{6}$

(114) $5 + \dfrac{5}{k} + \dfrac{1}{5k^2}$

(115) $2 + \dfrac{1}{x} + \dfrac{1}{6x^2}$

(116) $\dfrac{2n^3}{9} + n^2 + 2n$

(117) $\dfrac{a}{3} + 1 + \dfrac{3}{a}$

(118) $5p^5 + \dfrac{3p^4}{8} + 2p^3$

(119) $4x^2 + 3x + \dfrac{2}{5}$

(120) $3n + \dfrac{1}{8} + \dfrac{1}{2n}$

(121) $\dfrac{n}{4} + 5 + \dfrac{1}{n}$

(122) $4n + \dfrac{1}{3} + \dfrac{1}{n}$

(123) $2v^5 + 3v^4 + 5v^3$

(124) $k^2 + \dfrac{k}{4} + 1$

(125) $4 + \dfrac{1}{3v} + \dfrac{2}{v^2}$

(126) $\dfrac{x^2}{3} + \dfrac{x}{3} + 2$

www.math-knots.com

(127) $5 + \dfrac{1}{5x} + \dfrac{1}{5x^2}$

(128) $\dfrac{x^5}{2} + 5x^4 + \dfrac{x^3}{6}$

(129) $3 + \dfrac{5}{n} + \dfrac{1}{n^2}$

(130) $m + 1 + \dfrac{1}{m}$

(131) $m + 3 + \dfrac{6}{m-7}$

(132) $n - 10 - \dfrac{2}{n+6}$

(133) $b - 9 - \dfrac{9}{b+10}$

(134) $k - 3 - \dfrac{10}{k+10}$

(135) $v + 2 - \dfrac{8}{v-9}$

(136) $m + 10 + \dfrac{9}{m-2}$

(137) $b - 12 + \dfrac{8}{b+7}$

(138) $7k - 1 + \dfrac{4}{k+9}$

(139) $x + 6 + \dfrac{1}{x-1}$

(140) $n + 10 - \dfrac{8}{n+11}$

(141) $7r - 3 - \dfrac{2}{r-7}$

(142) $2a - 9 + \dfrac{9}{a-12}$

(143) $x - 12 - \dfrac{1}{x+3}$

(144) $n + 7 + \dfrac{4}{n+8}$

(145) $r - 1 + \dfrac{12}{r + 9}$

(146) $x - 12 - \dfrac{11}{x + 8}$

(147) $x + 8 + \dfrac{8}{x + 4}$

(148) $x - 5 + \dfrac{11}{x + 4}$

(149) $n - 2 - \dfrac{6}{n + 5}$

(150) $n - 11 - \dfrac{6}{n - 3}$

(151) $x - 1 + \dfrac{12}{x + 11}$

(152) $r + 7 - \dfrac{1}{r + 10}$

(153) $r + 6 - \dfrac{4}{r - 5}$

(154) $k + 6 + \dfrac{7}{k - 3}$

(155) $x + 4 + \dfrac{7}{x + 11}$

(156) $n^2 + 3n - 8$

(157) $6x^2 + 4x - 9$

(158) $5a^2 - 2a - 9$

(159) $k^2 + 6k + 4$

(160) $n^2 + 7n - 4$

(161) $v^2 - 7v + 1$

(162) $4v^2 + 2v + 5$

www.math-knots.com

(163) $r^2 + 3r + 5$

(164) $m^2 + 10m - 3$

(165) $n^2 + 10n - 4$

(166) $p^2 - 10p - 4$

(167) $k^2 - 3k + 5$

(168) $x^2 - 3x - 9$

(169) $x^2 + 8x - 2$

(170) $2x^2 - 7x - 1$

(171) $x^2 - 9x - 7$

(172) $10p^2 - 2p - 1$

(173) $3m^2 + 10m + 2$

(174) $k^2 - 5k + 9$

(175) $n^2 - 9n + 6$

(176) $4x^2 + 8x + 8$

(177) $m^2 + 10m - 3$

(178) $m^2 + 9m + 6$

(179) $r^2 + 6r - 3$

(180) $b^2 + 3b - 4$

(181) $21p^2 - 50p - 16$

(182) $24a^2 - 29a + 7$

(183) $7x^2 - 30x + 8$

(184) $49b^2 - 84b + 32$

(185) $20a^2 + 11a - 4$

(186) $3p^2 + 14p - 5$

(187) $5k^2 - 9k - 2$

(188) $2x^2 - 6x - 8$

(189) $8n^2 - 10n - 12$

(190) $20x^2 - 24x + 4$

(191) $16k^2 + 26k - 12$

(192) $2m^2 + 4m - 48$

(193) $4p^2 - 36$

(194) $14v^2 + 28v + 14$

(195) $25x^2 + 50x + 24$

(196) $20r^2 + 48r + 28$

(197) $3n^2 + 18n - 48$

(198) $25k^2 + 20k - 32$

161   www.math-knots.com

(199) $3a^2 - 3a - 6$

(200) $32x^2 - 56x + 24$

(201) $12a^2 - 28a + 8$

(202) $12n^2 + 46n + 40$

(203) $24x^2 - 6$

(204) $8a^2 + 4a - 12$

(205) $36x^2 + 30x + 4$

(206) $15n^3 + 45n^2 + 51n + 42$

(207) $2n^3 + 5n^2 + n + 12$

(208) $5x^3 - 2x^2 + 2x + 3$

(209) $48x^3 - 64x^2 + 50x - 25$

(210) $4x^3 + 14x^2 + 38x + 16$

(211) $12m^3 - 12m^2 - 9m + 6$

(212) $30a^3 - 14a^2 - 24a - 4$

(213) $35x^3 - 84x^2 + 91x - 42$

(214) $4a^3 + 8a^2 + a - 1$

(215) $18k^3 + 51k^2 - k - 42$

(216) $18n^3 - 3n^2 - 24n - 3$

(217) $5v^3 + 9v^2 - 7v + 1$

(218) $3x^3 + 22x^2 + 14x + 49$

(219) $18n^3 - 12n^2 - 52n + 48$

(220) $2n^3 + 8n^2 - 36n - 18$

(221) $8n^3 + 15n^2 + n + 6$

(222) $20k^3 - k^2 - 13k + 3$

(223) $8n^3 - 63n^2 - 56n - 6$

(224) $8n^3 - 43n^2 - 31n + 6$

(225) $12k^3 - 2k^2 + 8k + 6$

(226) $8m^3 + 33m^2 - 59m + 21$

(227) $48b^3 - 18b^2 + 11b - 28$

(228) $3n^3 - 4n - 16$

(229) $12p^3 - 21p^2 + 5p + 3$

(230) $20k^3 - 9k^2 - 38k - 15$

(231) $7r^3 - 15r^2 - 53r - 30$

(232) $48n^3 + 28n^2 + 58n + 48$

(233) $6x^3 - 9x^2 - 28x - 49$

(234) $12n^3 - 44n^2 + 29n + 8$

www.math-knots.com

(235) $5a^3 + 12a^2 + 47a + 56$

(236) $n^4 - 4n^3 - 32n^2 - 24n - 4$

(237) $15x^4 - 23x^3 + 12x^2 - 5x + 1$

(238) $15m^4 - 12m^3 + 6m^2 + 9m - 6$

(239) $5m^4 - 16m^3 + 33m^2 + 8m - 35$

(240) $24k^4 - 28k^3 - 14k^2 + 33k - 5$

(241) $14k^4 - 8k^3 - 2k^2 + 16k - 10$

(242) $3n^4 + 2n^3 - 7n^2 - 18n - 10$

(243) $42r^4 + 38r^3 - 25r^2 - 16r + 6$

(244) $18x^4 + 18x^3 + 19x^2 + x + 7$

(245) $7x^4 + 16x^3 - 57x^2 + 56x - 49$

(246) $42v^4 - 15v^3 - 86v^2 + 19v + 40$

(47) $24x^4 - 56x^3 + 2x^2 - 42x - 49$

(248) $12x^4 + 18x^3 + 23x^2 + x - 40$

(249) $24k^4 - 53k^3 + 12k^2 - 5k + 2$

(250) $28v^4 - 12v^3 - 90v^2 + 8v + 48$

(251) $6x^4 - 8x^3 + 30x^2 - 12x + 16$

(252) $42m^4 - 48m^3 - 84m^2 + 54m + 48$

(253) $24x^4 - 16x^3 + 36x^2 - 45x + 14$

(254) $14n^4 - 35n^3 - 36n^2 + 61n + 28$

(255) $56x^4 - 8x^3 - 74x^2 - 40x - 6$

(256) $9r^4 + 15r^3 - 12r^2 + 54r - 12$

(257) $40k^4 - 6k^3 - 53k^2 + 33k - 5$

(258) $24x^4 - 27x^3 + 54x^2 - 9x + 6$

(259) $24v^4 + 12v^3 - 80v^2 + 54v - 8$

(260) $7n^4 + 26n^3 - 56n^2 + 18n - 7$

(261) $5a(9a + 10)$

(262) $-3a^2(3 + a)$

(263) $7x^2(2x^2 + 3)$

(264) $-4b(8b + 9)$

(265) $10(-5x + 2)$

(266) $6a(7a^4 + 4)$

(267) $6(3x + 8)$

(268) $7x^2(-5x + 2)$

(269) $3(-1 + 3a^2)$

(270) $8(-8b + 3)$

(271) $9k(k^7 + 2)$

(272) $4m^5(2m + 9)$

(273) $6a^4(2a - 3)$

(274) $7p^3(-3p + 2)$

(275) $5(k^4 - 8)$

(276) $8x^2(-7x^3 + 6)$

(277) $9r(-5r + 4)$

(278) $4x(2x^2 - 5)$

(279) $2p^2(5p^2 + 4)$

(280) $5b^2(-b^3 + 2)$

(281) $8(3v^7 - 8)$

(282) $7(4x^2 - 5)$

(283) $9(7m^3 + 1)$

(284) $10r^2(10 - 7r)$

(285) $2(m^4 + 2)$

(286) $x^2 + 6xy + 9y^2$

(287) $a^2 - 28ab + 196b^2$

(288) $n^2 + 90nm + 2025m^2$

(289) $y^2 - 68yx + 1156x^2$

(290) $u^2 - 18uv + 81v^2$

(291) $x^2 + 62xy + 961y^2$

(292) $a^2 - 50ab + 625b^2$

(293) $u^2 + 82uv + 1681v^2$

(294) $x^2 + 78xy + 1521y^2$

(295) $y^2 + 44yx + 484x^2$

(296) $v^2 - 58vu + 841u^2$

(297) $u^2 - 84uv + 1764v^2$

(298) $a^2 - 4ab + 4b^2$

(299) $u^2 + 46uv + 529v^2$

(300) $x^2 + 94xy + 2209y^2$

(301) $y^2 + 92yx + 2116x^2$

(302) $n^2 - 56nm + 784m^2$

(303) $a^2 - 62ab + 961b^2$

(304) $m^2 + 12mn + 36n^2$

(305) $x^2 - 6xy + 9y^2$

(306) $n^2 + 4nm + 4m^2$

(307) $a^2 + 100ab + 2500b^2$

(308) $a^2 + 98ab + 2401b^2$

(309) $x^2 + 66xy + 1089y^2$

(310) $v^2 - 52vu + 676u^2$

(311) $m^2 + 28mn + 196n^2$

(312) $u^2 + 96uv + 2304v^2$

(313) $u^2 - 40uv + 400v^2$

(314) $x^2 - 86xy + 1849y^2$

(315) $y^2 - 22yx + 121x^2$

(316) $x^2 - 100y^2$

(317) $x^2 - 49y^2$

(318) $x^2 - 36y^2$

(319) $x^2 - 144y^2$

(320) $m^2 - 196n^2$

(321) $u^2 - 121v^2$

(322) $u^2 - 81v^2$

(323) $m^2 - 169n^2$

(324) $b^2 - 121a^2$

www.math-knots.com

(325) $x^2 - 9y^2$

(326) $x^2 - 25y^2$

(327) $y^2 - 16x^2$

(328) $u^2 - 4v^2$

(329) $n^2 - 16m^2$

(330) $x^2 - 100y^2$

(331) $x^2 - 81y^2$

(332) $m^2 - n^2$

(333) $x^2 - 169y^2$

(334) $b^2 - 144a^2$

(335) $u^2 - 25v^2$

(336) $y^2 - 49x^2$

(337) $x^2 - y^2$

(338) $u^2 - 64v^2$

(339) $v^2 - 36u^2$

(340) $u^2 - 9v^2$

(341) $7(6x + 11y)(2x - 3y)$

(342) $(4u + 3v)(3u + 8v)$

(343) $8(x + 7y)(9x + 10y)$

(344) $2x(7x + 13y)$

(345) $(x - 2y)(9x + 2y)$

(346) $5(x - 11y)(8x + 7y)$

(347) $(x + 8y)(8x + 11y)$

(348) $(a + 7b)(10a + b)$

(349) $(2x - 3y)(7x + 12y)$

(350) $(5u + v)(2u - v)$

(351) $(x - 6y)(8x + 5y)$

(352) $(a + 8b)(14a - 5b)$

(353) $(a + 4b)(9a - 4b)$

(354) $4(3x + 7y)(3x + 10y)$

(355) $4(a - 8b)(4a - 11b)$

(356) $(x + y)(9x - 10y)$

(357) $2(5x - 3y)(2x - 3y)$

(358) $2(x - 10y)(6x + y)$

(359) $2(u + 8v)(6u - 11v)$

(360) $(x - 3y)(4x - y)$

(361) $(4x - 3y)(2x - 5y)$

(362) $(3a + 10b)(3a - 8b)$

(363) $(6x + 7y)(2x - y)$

(364) $(x + 2y)(9x + 10y)$

(365) $(x + 2y)(4x + 11y)$

(366) $(2u - 11v)(5u - 3v)$

(367) $4(3x + 7y)(3x - 10y)$

(368) $2(2x - 3y)(7x + 10y)$

(369) $2m(3m + 8n)$

(370) $5(x + 10y)(9x - 4y)$

(371) $(n - 9)(9n + 4)$

(372) $(3m + 11)(2m - 9)$

(373) $7(x - 1)(10x - 9)$

(374) $(x - 8)(9x - 10)$

(375) $2(2x - 1)(3x - 8)$

(376) $(n - 9)(8n + 9)$

(377) $(k + 4)(9k + 2)$

(378) $(p - 3)(10p - 9)$

www.math-knots.com

(379) $5(x - 10)(9x + 2)$

(380) $(v + 9)(10v + 9)$

(381) $(2n + 7)(5n + 12)$

(382) $(k + 10)(4k + 3)$

(383) $(x + 4)(12x + 7)$

(384) $3(v + 10)(4v + 3)$

(385) $(n - 1)(6n + 5)$

(386) $(n + 10)(4n - 9)$

(387) $2(a - 9)(6a + 5)$

(388) $(a - 10)(9a - 2)$

(389) $(3n + 2)(2n - 1)$

(390) $6(b + 2)(9b - 1)$

(391) $6(3b + 10)(3b + 7)$

(392) $5(5n + 6)(2n - 3)$

(393) $2(2n + 9)(5n + 6)$

(394) $4(3r - 10)(2r - 5)$

(395) $6(p - 6)(10p + 9)$

(396) $6(2x + 5)(3x - 4)$

www.math-knots.com

(397) $2(2p - 3)(4p + 7)$

(398) $5(r + 4)(9r + 1)$

(399) $3(m + 4)(4m - 1)$

(400) $3(x - 1)(9x - 4)$

(401) $3(a + 3)(6a + 5)$

(402) $2(2x + 9)(3x - 10)$

(403) $2(2n - 1)(3n - 10)$

(404) $6(n - 1)(9n + 8)$

(405) $3(v + 6)(9v - 4)$

(406) $2(3x + 4)(2x - 9)$

(407) $4(5x - 4)(2x + 9)$

(408) $6(v - 2)(8v - 5)$

(409) $2(n - 10)(9n + 4)$

(410) $5(3p - 10)(2p + 1)$

(411) $4(u + 3v)(u - 3v)$

(412) $2(3m + 4n)(3m - 4n)$

(413) $2(3u + 2v)(3u - 2v)$

(414) $4(3x + 4y)(3x - 4y)$

(415) $5(4x + 5y)(4x - 5y)$

(416) $4(5x + 2y)(5x - 2y)$

(417) $4(a + 4b)(a - 4b)$

(418) $3(3x + 5y)(3x - 5y)$

(419) $4(5x + y)(5x - y)$

(420) $4(x + 5y)(x - 5y)$

(421) $3(x + 2y)(x - 2y)$

(422) $2(2x + y)(2x - y)$

(423) $5(x + 2y)(x - 2y)$

(424) $3(5x + 2y)(5x - 2y)$

(425) $5(3x + y)(3x - y)$

(426) $3(5x + 4y)(5x - 4y)$

(427) $4(4u + 5v)(4u - 5v)$

(428) $5(x + y)(x - y)$

(429) $5(u + 5v)(u - 5v)$

(430) $2(x + 4y)(x - 4y)$

(431) $12(7x - 11y)^2$

(432) $5(14x - y)^2$

(433) $4(8x - 9y)^2$

(434) $13(13x - 5y)^2$

(435) $11(9x + 10y)^2$

(436) $11(4x - y)^2$

(437) $12(6x - 13y)^2$

(438) $12(7x + 3y)^2$

(439) $3(x - 7y)^2$

(440) $11(3x + 2y)^2$

(441) $12(u - 3v)^2$

(442) $11(4u + 9v)^2$

(443) $13(5x + 14y)^2$

(444) $13(5x + 12y)^2$

(445) $9(10x - 3y)^2$

(446) $7(9x + 8y)^2$

(447) $11(x - 7y)^2$

(448) $12(4a + 3b)^2$

(449) $2(7x - 5y)^2$

(450) $13(13x + 10y)^2$

www.math-knots.com

(451) $9(10x + 13y)^2$

(452) $3(5a - 6b)^2$

(453) $7(13u - 7v)^2$

(454) $9(5x - 4y)^2$

(455) $14(12x + 7y)^2$

(456) $(7b - 5x)(8z + 7c)$

(457) $(x + 6y)(10c + f)$

(458) $(10b - 7x)(9z^2 - 5h)$

(459) $(7a - 6y)(8z^2 + h)$

(460) $(x - y)(2u^2 + 3v)$

(461) $(5x^2 + 4y)(2u + v)$

(462) $(7x - 3y)(2w + 5f)$

(463) $(7b + 6x)(10z^2 - 7h)$

(464) $(4x + y)(3c - 4d)$

(465) $(7b - 2x)(3c - f)(3c + f)$

(466) $(9a + 8b)(9c + 10f)$

(467) $(5x + 7y)(3c - 4f)$

(468) $(a - 9b)(7c + 8f)$

(469) $\left(7a - 10y\right)\left(9h^2 - 8f\right)$

(470) $\left(3p - 7q\right)\left(9u + 7v\right)$

(471) $\left(10a + 9y\right)\left(7h - 2f\right)$

(472) $\left(3m + 7n\right)\left(h + 2f\right)$

(473) $\left(7x + 6y\right)\left(10c + 9f\right)$

(474) $\left(8x - 9y\right)\left(3z + 8h\right)$

(475) $\left(5b - 3x\right)\left(z - 3h\right)$

(476) $\left(3a - x\right)\left(u - 7v\right)$

(477) $\left(7a + 6y\right)\left(4c - 5d\right)$

(478) $\left(3p^2 - 10q\right)\left(3c - 8d^2\right)$

(479) $\left(7m + n\right)\left(8h + f\right)$

(480) $\left(4a^2 - x\right)\left(2z - 5c\right)$

(481) $\left(5a - b\right)\left(10h + 7f\right)$

(482) $\left(a^2 + 3y\right)\left(5z + 4h\right)$

(483) $\left(b + x\right)\left(7z - 6h\right)$

(484) $\left(9a - 10b\right)\left(10h + 3f\right)$

(485) $\left(10b + 9x\right)\left(h + f\right)$

www.math-knots.com

(486)

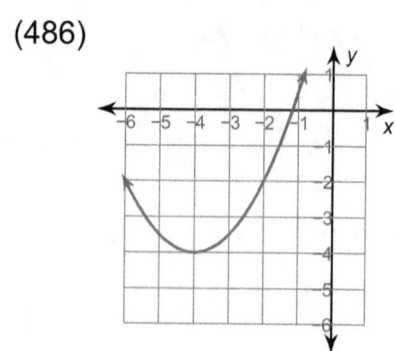

(487)

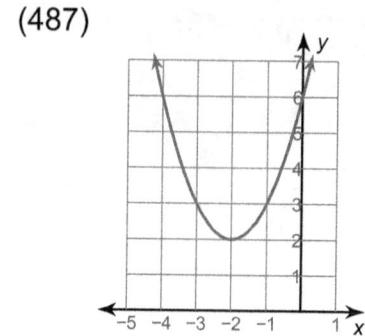

(488)

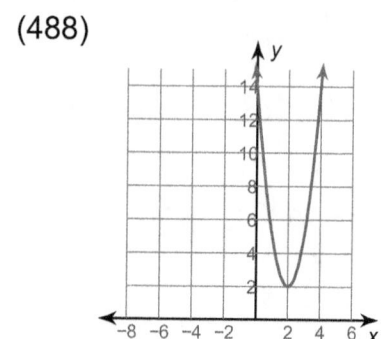

(489)

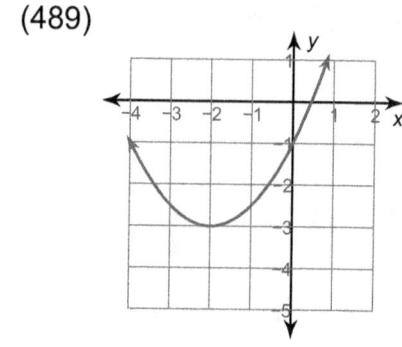

(490)

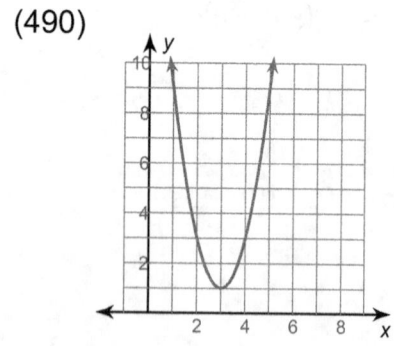

(491)

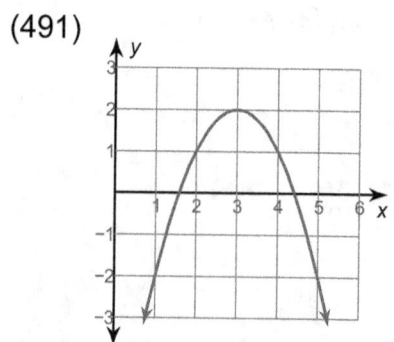

www.math-knots.com

(492)

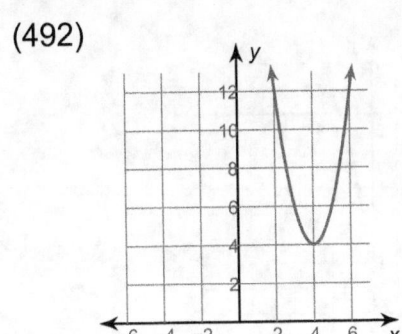

(493)

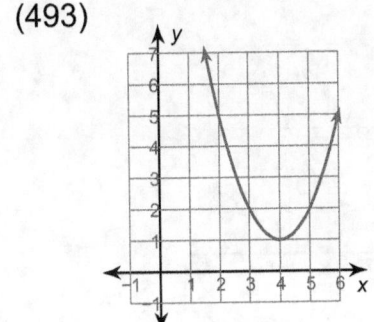

(494)

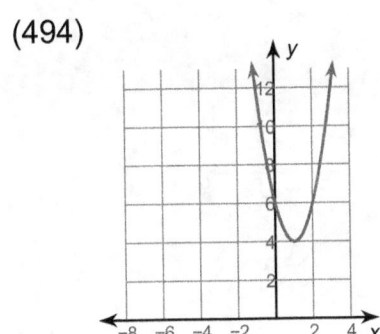

(495)

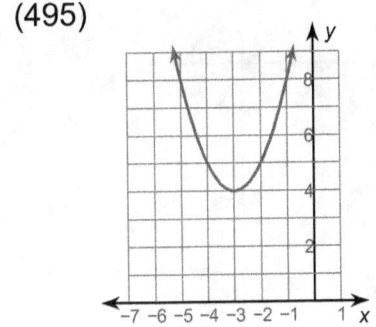

(496)

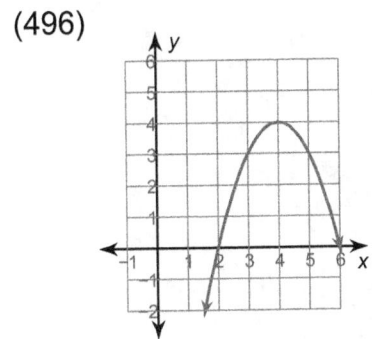

(497)

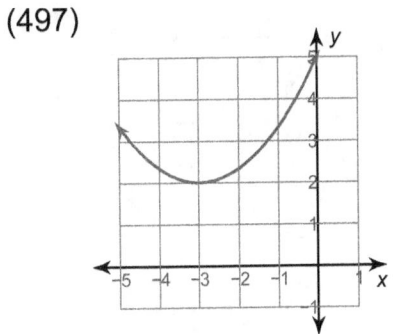

www.math-knots.com

(498)

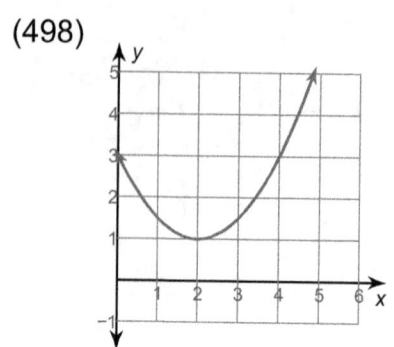

(499)

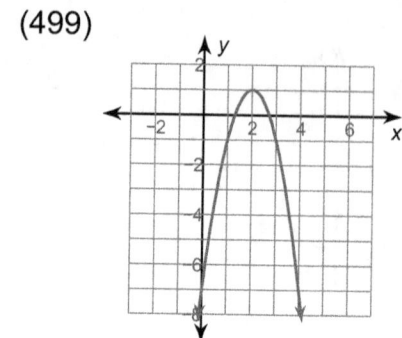

(500)

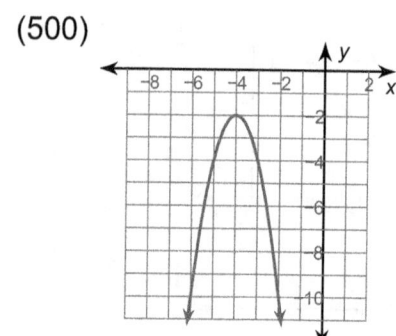

(501)

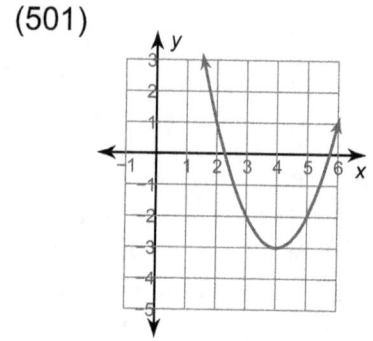

(502)

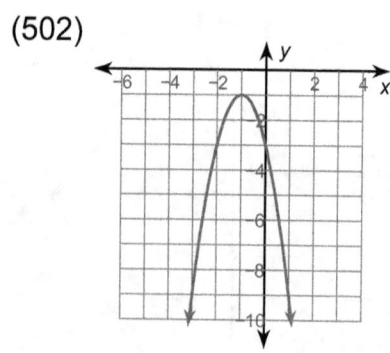

(503)

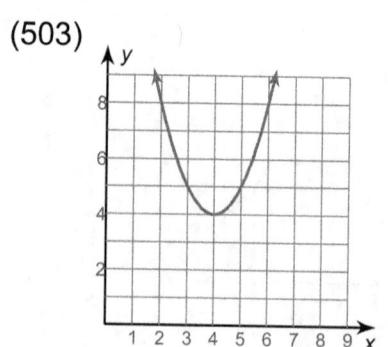

(504)

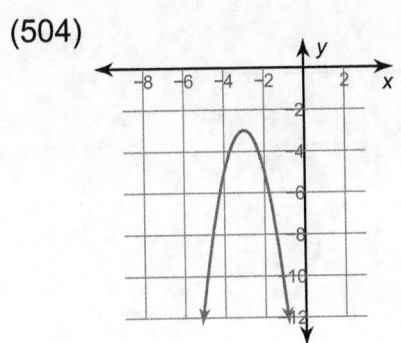

(505)

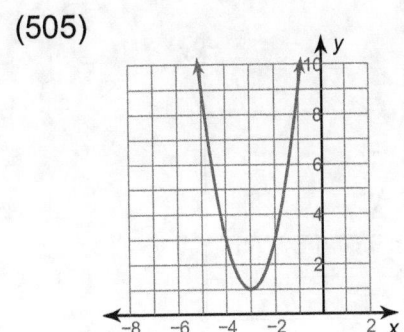

(506)

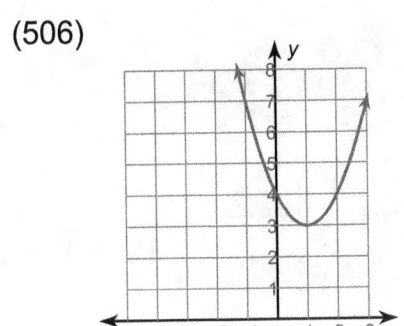

(507)

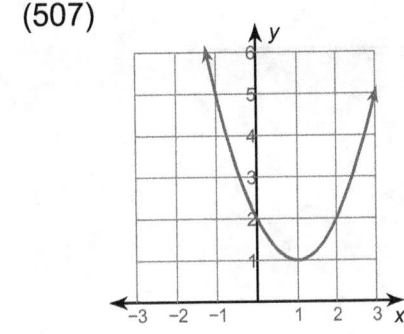

(508)

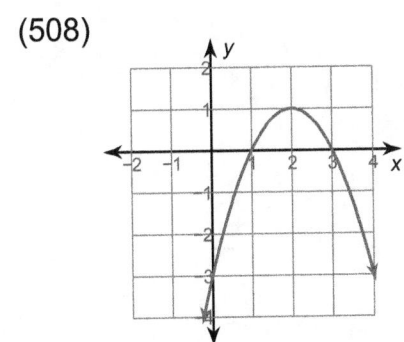

(509)

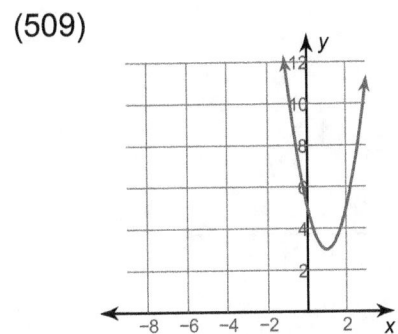

www.math-knots.com

(510)

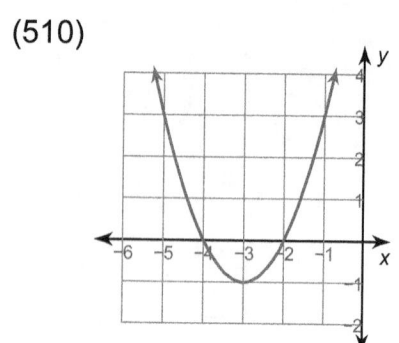

(511)

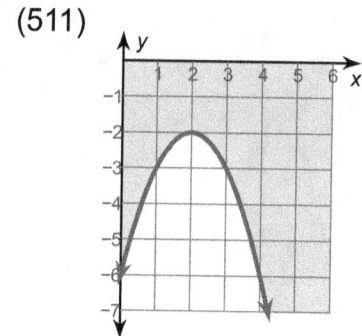

(512)

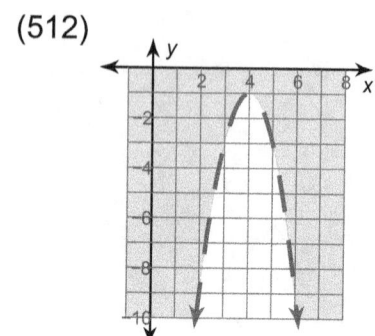

(513)

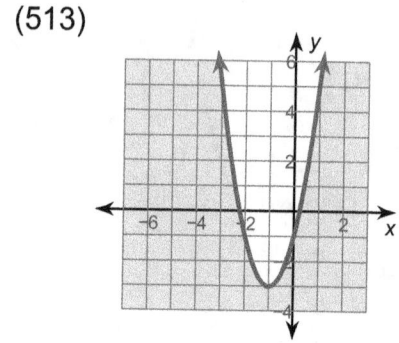

(514)

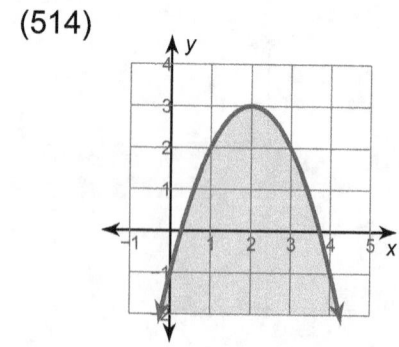

(515)

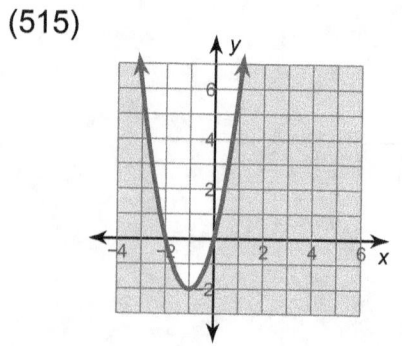

(516)

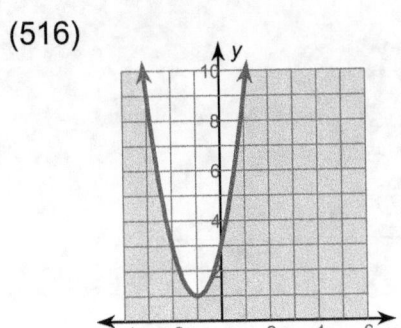

(517)

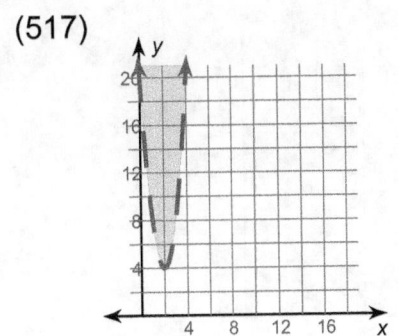

(518)

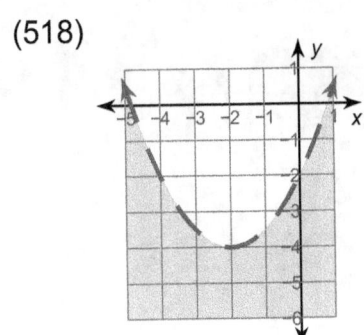

(519)

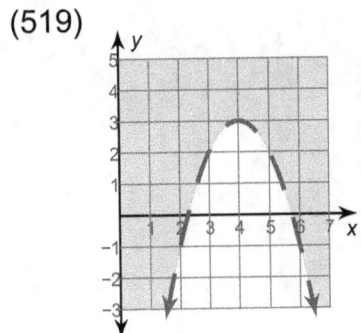

(520)

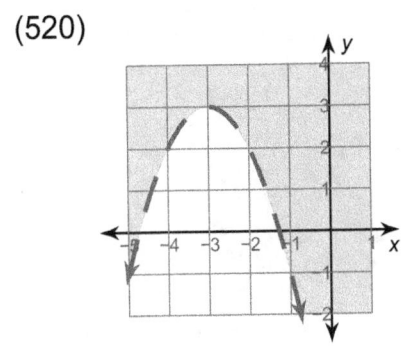

(521)

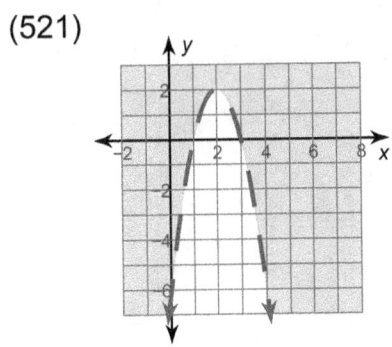

www.math-knots.com

(522)

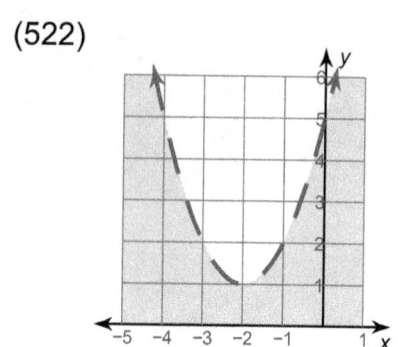

(523)

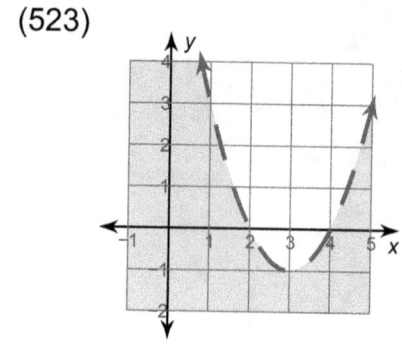

(524)

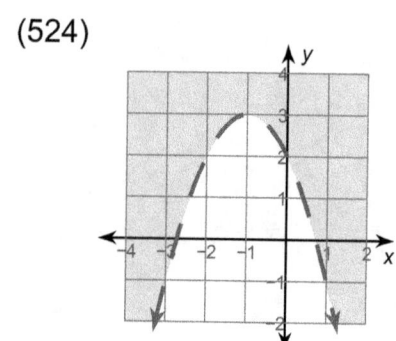

(525)

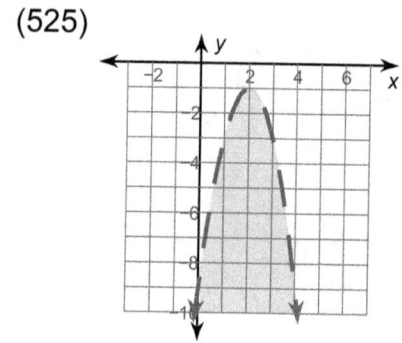

(526)

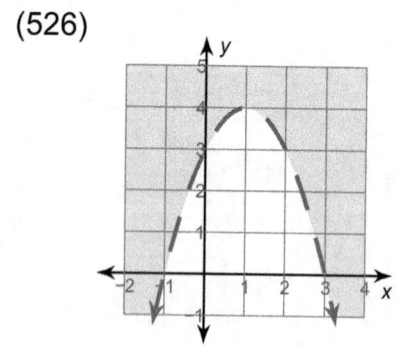

(527)

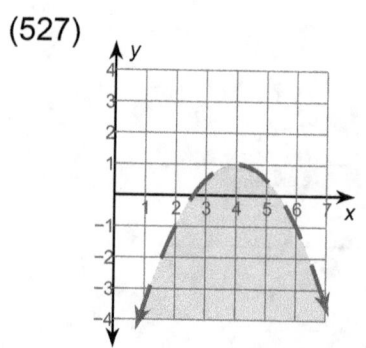

(528)

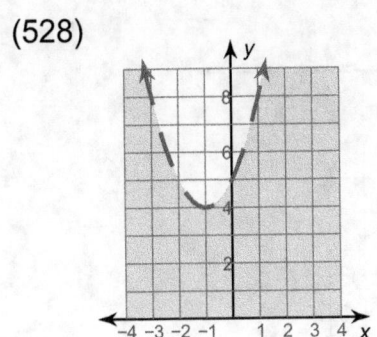

(529)

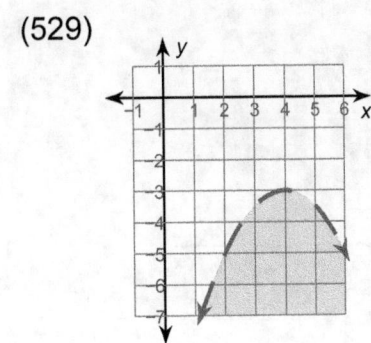

(530)

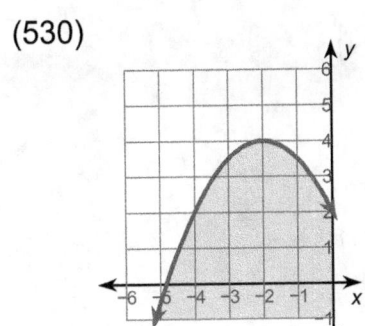

(531)

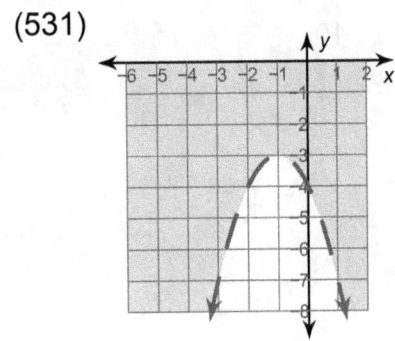

(532)

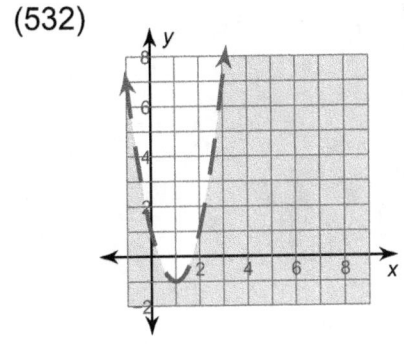

(533)

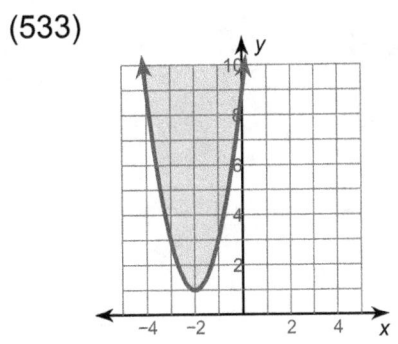

www.math-knots.com

(534)

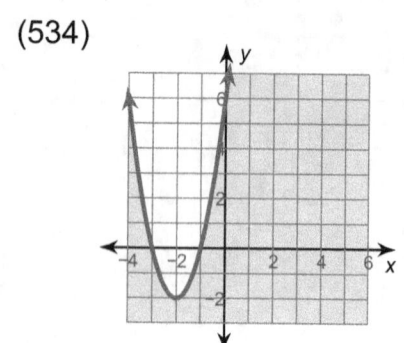

(535)

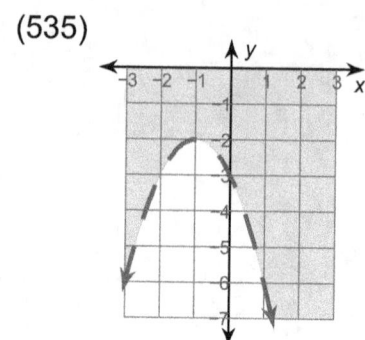

(536) $\{10, -10\}$

(537) $\{4, -4\}$

(538) $\{2, -2\}$

(539) $\{1, -1\}$

(440) $\{7, -7\}$

(541) $\{7, -7\}$

(542) $\{2, -2\}$

(543) $\{8, -8\}$

(544) $\{5, -5\}$

(545) $\left\{1\frac{1}{9}, -1\frac{1}{9}\right\}$

(546) $\{2, -2\}$

(547) $\left\{\frac{5}{8}, -\frac{5}{8}\right\}$

(548) $\{4, -4\}$

(549) $\{3, -3\}$

(550) $\{10, -10\}$

(551) $\left\{\dfrac{1}{2}, -\dfrac{1}{2}\right\}$

(552) $\{10, -10\}$

(553) $\{3, -3\}$

(554) $\{7, -7\}$

(555) $\left\{\dfrac{7}{8}, -\dfrac{7}{8}\right\}$

(556) $\{10, -10\}$

(557) $\{1, -1\}$

(558) $\{1, -1\}$

(559) $\{2, -2\}$

(560) $\left\{\dfrac{3}{7}, -\dfrac{3}{7}\right\}$

(561) $\{4, 9\}$

(562) $\{-5, -7\}$

(563) $\{11, -6\}$

(564) $\{-4, -1\}$

(565) $\{7, 0\}$

www.math-knots.com

(566) $\{-7, 7\}$

(567) $\{-12, -7\}$

(568) $\{-3, 6\}$

(569) $\{12, -1\}$

(570) $\{-5, 7\}$

(571) $\{-10, 11\}$

(572) $\{1, 9\}$

(573) $\{-9, -7\}$

(574) $\{9, -11\}$

(575) $\{-6, 0\}$

(576) $\{12, 0\}$

(577) $\{10, -12\}$

(578) $\{2, 4\}$

(579) $\{8, 0\}$

(580) $\{-5, 0\}$

(581) $\{4, 0\}$

(582) $\{-8, 7\}$

(583) $\{2, -9\}$

(584)   {−5, −11}

(585)   {−12, −11}

(586)   {−15, −20}

(587)   {9, −7}

(588)   {20, 0}

(589)   {11, −20}

(590)   {−14, −5}

(591)   {12, −10}

(592)   {−20, 1}

(593)   {11, 10}

(594)   {−8, 0}

(595)   {20, −20}

(596)   {−10, −20}

(597)   {−16, 13}

(598)   {6, −19}

(599)   {10, −9}

(600)   {17, −18}

(601)   {2, 11}

www.math-knots.com

(602)   {4, 16}

(603)   {7, 13}

(604)   {−9, 15}

(605)   {13, 10}

(606)   −32; two imaginary solutions

(607)   0; one real solution

(608)   0; one real solution

(609)   −3; two imaginary solutions

(610)   −179; two imaginary solutions

(611)   0; one real solution

(612)   289; two real solutions

(613)   0; one real solution

(614)   0; one real solution

(615)   −92; two imaginary solutions

(616)   225; two real solutions

(617)   4; two real solutions

(618)   0; one real solution

(619)   −12; two imaginary solutions

(620)   36 ; two real solutions

(621)   0 ; one real solution

(622)   0 ; one real solution

(623)   36 ; two real solutions

(624)   144 ; two real solutions

(625)   0 ; one real solution

(626)   −215 ; two imaginary solutions

(627)   25 ; two real solutions

(628)   −279 ; two imaginary solutions

(629)   0 ; one real solution

(630)   0 ; one real solution

(631) $\dfrac{16}{25}$

(632)   100

(633) $\dfrac{4}{169}$

(634)   196

(635)   256

(636) $\dfrac{121}{1156}$

(637) $\dfrac{9}{4}$

(638) $\dfrac{225}{676}$

(639) $\dfrac{16129}{576}$

(640) 256

(641) 324

(642) $\dfrac{441}{4}$

(643) $\dfrac{4761}{1600}$

(644) 4

(645) 289

(646) $\dfrac{169}{36}$

(647) 81

(648) $\dfrac{49}{4}$

(649) 441

(650) 1

(651) $\dfrac{225}{4}$

(652) 441

(653) 36

(654) $\dfrac{361}{4}$

(655) 16

www.math-knots.com

(656) $\dfrac{289}{4}$

(657) $\dfrac{1225}{1296}$

(658) $\dfrac{25}{4}$

(659) 361

(660) $\dfrac{169}{1024}$

(661) 196

(662) $\dfrac{225}{4}$

(663) $\dfrac{121}{4}$

(664) 289

(665) $\dfrac{25}{36}$

(666) 400

(667) $\dfrac{625}{256}$

(668) 81

(669) $\dfrac{289}{4}$

(670) $\dfrac{121}{4}$

(671) 169

(672) $\dfrac{9}{4}$

(673) $\dfrac{25}{4}$

(674)   36

(675)   64

(676)   169

(677)   49

(678)   9

(679) $\dfrac{1}{25}$

(680) $\dfrac{49}{4}$

(681) $\{6.211, -8.211\}$

(682) $\{-3, -13\}$

(683) $\{10, -2\}$

(684) $\{8, 2\}$

(685) $\{0.724, -20.724\}$

(686) $\{-2, -4\}$

(687) $\{-2, -8\}$

(688) $\{5, -13\}$

(689) $\{3.472, -5.472\}$

(690) $\{11.592, -7.592\}$

(691) $\{1, -9\}$

(692) $\{6.775, -10.775\}$　　　(693) $\{8, -10\}$

(694) $\{13, 5\}$　　　(695) $\{9, 5\}$

(696) $\{2, -4\}$　　　(697) $\{-2, -14\}$

(698) $\{5, -3\}$　　　(699) $\{-4.172, -9.828\}$

(700) $\{6, -2\}$　　　(701) $\{20, -4\}$

(702) $\{13, -3\}$　　　(703) $\{11.928, -1.928\}$

(704) $\{19, 1\}$　　　(705) $\{13, -7\}$

(706) $14p\sqrt{2}$　　　(707) $3x\sqrt{6x}$

(708) $6p\sqrt{2p}$　　　(709) $5v^2\sqrt{2}$

www.math-knots.com

(710) $5\sqrt{6n}$

(711) $8x\sqrt{2x}$

(712) $8m\sqrt{3}$

(713) $6b^2\sqrt{2}$

(714) $4b\sqrt{5b}$

(715) $2n\sqrt{2}$

(716) $6\sqrt{2x}$

(717) $8\sqrt{2x}$

(718) $14m\sqrt{m}$

(719) $7v\sqrt{7}$

(720) $7a\sqrt{2}$

(721) $4p^2\sqrt{7}$

(722) $4x\sqrt{2}$

(723) $2\sqrt{6x}$

(724) $8x\sqrt{7}$

(725) $2p^2\sqrt{5}$

(726) $5x^2y^2\sqrt{10}$

(727) $12xy\sqrt{7}$

(728) $12m^2n^3\sqrt{mn}$

(729) $14u^3v^3\sqrt{3}$

(730) $24v^2u^3\sqrt{2u}$

(731) $8u^3\sqrt{3uv}$

(732) $8b\sqrt{3ab}$

(733) $6x^3y\sqrt{6}$

(734) $8x^2y^2\sqrt{x}$

(735) $10u^3v\sqrt{15}$

(736) $4ab^3\sqrt{6a}$

(737) $4u^3v^3\sqrt{3v}$

(738) $14y^3\sqrt{7xy}$

(739) $5x^2y^3\sqrt{6xy}$

(740) $13u\sqrt{10v}$

(741) $3\sqrt{6xy}$

(742) $2mn^3\sqrt{6n}$

(743) $14xy\sqrt{7x}$

(744) $9mn^3\sqrt{3}$

(745) $28y^2x\sqrt{x}$

(746) $6p^2q^2\sqrt{2r}$

(747) $4z^2y\sqrt{5x}$

(748) $7k\sqrt{2hjk}$

(749) $12m^2p^2q$

(750) $6ab\sqrt{7abc}$

(751) $2q^2mp\sqrt{5m}$

(752) $5m^2p\sqrt{2pq}$

(753) $12xyz\sqrt{z}$

(754) $14y^2xz\sqrt{z}$

(755) $4x^2y^2\sqrt{z}$

(756) $5r^2pq\sqrt{5p}$

(757) $6xy\sqrt{6yz}$

(758) $3m\sqrt{3mnp}$

(759) $6x^2z\sqrt{7y}$

(760) $3q\sqrt{5pr}$

(761) $10xy\sqrt{xz}$

(762) $14yz\sqrt{x}$

(763) $4pq\sqrt{6mq}$

(764) $\quad 2p^2m\sqrt{6n}$

(765) $\quad 16y^2xz$

(766) $\quad -6\sqrt{3}$

(767) $\quad 2\sqrt{5}$

(768) $\quad 8\sqrt{6} - \sqrt{5}$

(769) $\quad -15\sqrt{3} - \sqrt{6}$

(770) $\quad 3\sqrt{5} - 2\sqrt{6}$

(771) $\quad -5\sqrt{6}$

(772) $\quad 3\sqrt{2} + \sqrt{3}$

(773) $\quad 3\sqrt{2} - \sqrt{3}$

(774) $\quad 5\sqrt{5} - 4\sqrt{2}$

(775) $\quad -18\sqrt{2}$

(776) $\quad -11\sqrt{5} - 2\sqrt{6}$

(777) $\quad -6\sqrt{3} - 15\sqrt{2}$

(778) $\quad -3\sqrt{3} + 8\sqrt{5}$

(779) $\quad 4\sqrt{2} + 3\sqrt{5}$

(780) $\quad -11\sqrt{3} - \sqrt{6}$

(781) $\quad 6\sqrt{5} + 3\sqrt{2}$

(782) $9\sqrt{5} + 3\sqrt{2}$

(783) $-2\sqrt{3} + 6\sqrt{2}$

(784) $-3\sqrt{5}$

(785) $-10\sqrt{5}$

(786) $-2\sqrt{5}$

(787) $-6\sqrt{5}$

(788) $-2\sqrt{3}$

(789) $8\sqrt{2} - 4\sqrt{3}$

(790) $11\sqrt{6} - 6\sqrt{2}$

(791) $\sqrt{30} + 3\sqrt{5}$

(792) $2\sqrt{15} + 3\sqrt{5}$

(793) $-8\sqrt{15} + 5\sqrt{2}$

(794) $2\sqrt{15} + 2\sqrt{3}$

(795) $20\sqrt{2} + 32\sqrt{5}$

(796) $5\sqrt{15} + 15\sqrt{10}$

(797) $5\sqrt{6} + 5\sqrt{15}$

(798) $10\sqrt{2} + 2\sqrt{30}$

(799) $3\sqrt{30} + 15\sqrt{3}$

(800) $3\sqrt{10} + 2\sqrt{5}$

(801) $3\sqrt{2} + \sqrt{30}$

(802) $2\sqrt{6} - 6$

(803) $4\sqrt{10} + 2\sqrt{5}$

(804) $\sqrt{30} - 20\sqrt{6}$

(805) $-20 + 2\sqrt{5}$

(806) $20\sqrt{5} + 25\sqrt{2}$

(807) $3\sqrt{5} + 5\sqrt{15}$

(808) $2\sqrt{5} + \sqrt{6}$

(809) $-6\sqrt{2} + \sqrt{30}$

(810) $6\sqrt{3} - 3\sqrt{30}$

(811) $-\dfrac{\sqrt{2}}{5}$

(812) $\dfrac{2\sqrt{5}}{5}$

(813) $\dfrac{\sqrt{15}}{5}$

(814) $-\dfrac{4\sqrt{3}}{3}$

(815) $\dfrac{\sqrt{15}}{6}$

(816) $\dfrac{4\sqrt{15}}{3}$

(817) $\dfrac{\sqrt{15}}{15}$

(18) $\dfrac{\sqrt{10}}{5}$

(819) $\dfrac{5\sqrt{3}}{3}$

(820) $\dfrac{\sqrt{10}}{2}$

(821) $\dfrac{3\sqrt{2}-2\sqrt{3}}{6}$

(822) $\dfrac{10-\sqrt{10}}{10}$

(823) $\dfrac{5\sqrt{7}+\sqrt{21}}{7}$

(824) $\dfrac{-2\sqrt{3}+\sqrt{6}}{6}$

(825) $\dfrac{-2\sqrt{5}+25}{5}$

(826) $\dfrac{\sqrt{3}+2\sqrt{6}}{3}$

(827) $\dfrac{3\sqrt{2}-\sqrt{6}}{4}$

(828) $\dfrac{-4\sqrt{19}-\sqrt{57}}{57}$

(829) $\dfrac{4\sqrt{13}+4\sqrt{65}}{13}$

(830) $\dfrac{2\sqrt{14}+\sqrt{7}}{28}$

(831) $\dfrac{4\sqrt{11}-\sqrt{22}}{55}$

(832) $\dfrac{\sqrt{15}-20}{15}$

(833) $\dfrac{3\sqrt{5}-3\sqrt{15}}{10}$

(834) $\dfrac{\sqrt{65}+3\sqrt{39}}{13}$

(835) $\dfrac{-5\sqrt{5}+3\sqrt{15}}{15}$

202   www.math-knots.com

(836) $\dfrac{-5\sqrt{2} - 3\sqrt{6} - 10\sqrt{5} - 6\sqrt{15}}{2}$

(837) $\dfrac{\sqrt{2} - 4\sqrt{5} + 3\sqrt{10} - 60}{78}$

(838) $\dfrac{20\sqrt{5} + 5\sqrt{15} + 4\sqrt{3} + 3}{13}$

(839) $12 + 8\sqrt{2} - 3\sqrt{3} - 2\sqrt{6}$

(840) $3 - \sqrt{3}$

(841) $\dfrac{3 - \sqrt{2}}{2}$

(842) $\dfrac{-5 - \sqrt{3}}{5}$

(843) $\dfrac{28 + 22\sqrt{2}}{23}$

(844) $\dfrac{-3 - 3\sqrt{2} + 2\sqrt{5} + 2\sqrt{10}}{5}$

(845) $-3 - 3\sqrt{2} + \sqrt{5} + \sqrt{10}$

(846) $\dfrac{10\sqrt{5} - 2\sqrt{3} + 5\sqrt{15} - 3}{122}$

(847) $\dfrac{20\sqrt{3} - 4\sqrt{2} - 10\sqrt{6} + 4}{73}$

(848) $\dfrac{-2 - 2\sqrt{5} - \sqrt{2} - \sqrt{10}}{20}$

(849) $\dfrac{12 - 4\sqrt{3} + 9\sqrt{2} - 3\sqrt{6}}{6}$

(850) $\dfrac{-15\sqrt{3} + 25 - 15\sqrt{6} + 25\sqrt{2}}{2}$

(851) $\dfrac{-6 + 2\sqrt{3} - 3\sqrt{5} + \sqrt{15}}{6}$

(852) $\dfrac{-20 + 4\sqrt{3}}{11}$

(853) $\dfrac{15 + 5\sqrt{2} + 12\sqrt{5} + 4\sqrt{10}}{7}$

(854) $\dfrac{-20 - 15\sqrt{2}}{2}$

(855) $\dfrac{-5 - 5\sqrt{5} - \sqrt{2} - \sqrt{10}}{20}$

(856) $\dfrac{-37\sqrt{2} + 50}{34}$

(857) $\dfrac{20 + 4\sqrt{2} - 5\sqrt{5} - \sqrt{10}}{23}$

(858) $\dfrac{25 - 5\sqrt{3} - 25\sqrt{5} + 5\sqrt{15}}{22}$

(859) $\dfrac{-12 - 4\sqrt{2} + 3\sqrt{3} + \sqrt{6}}{7}$

(860) $\dfrac{-5\sqrt{5} + 5\sqrt{2} + 15 - 3\sqrt{10}}{3}$

(861) $\dfrac{-87 + 27\sqrt{5}}{109}$

(862) $\dfrac{5 + 10\sqrt{5} - 4\sqrt{3} - 8\sqrt{15}}{19}$

(863) -2

(864) $\dfrac{1 - 2\sqrt{2}}{7}$

(865) $\dfrac{4 + 10\sqrt{2} - 2\sqrt{3} - 5\sqrt{6}}{46}$

(866) $\{-6\}$

(867) $\{160\}$

(868) $\{1\}$

(869) $\{0\}$

(870) $\{1\}$

(871) $\{3\}$

www.math-knots.com

(872) {75}

(873) {7}

(874) {3}

(875) {0}

(876) {2}

(877) {−1}

(878) {2}

(879) {9}

(880) {59}

(881) {1}

(882) {3}

(883) {7}

(884) {1}

(885) {18}

(886) {9}

(887) {3}

(888) {2}

(889) {3}

(890) $\{15\}$

(891) $\{5\}$

(892) $\{4, 9\}$

(893) $\{5, 1\}$

(894) $\{6, 1\}$

(895) $\{2, 8\}$

(896) $\{0, 8\}$

(897) $\{7\}$

(898) $\{9, 4\}$

(899) $\{8\}$

(900) $\{1\}$

(901) $\{8, 7\}$

(902) $\{9\}$

(903) $\{-1\}$

(904) $\{9, 7\}$

(905) $\{8\}$

(906) $\{-1, 1\}$

(907) $\{7\}$

(908) $\{6\}$

(909) $\{3\}$

(910) $\{1\}$

(911) $\dfrac{2n(n-2)}{3}$

(912) -1

(913) $\dfrac{(b+8)^2}{b+10}$

(914) $-6n$

(915) $-\dfrac{(n-4)(n+8)}{2n^2}$

(916) $\dfrac{2(n-2)}{n+3}$

(917) $n-4$

(918) $\dfrac{x+3}{7}$

(919) $-\dfrac{25p}{4}$

(920) $\dfrac{8k^2}{k+10}$

(921) $\dfrac{-v+2}{5}$

(922) 8

(923) $-\dfrac{4}{b+2}$

(924) $\dfrac{(p+5)(p+4)}{6p}$

(925) $\dfrac{r+9}{(r+5)^2}$

www.math-knots.com

(926) $\dfrac{r+1}{2}$

(927) $-\dfrac{3b}{7}$

(928) $\dfrac{3n}{8(n-1)}$

(929) $r+5$

(930) $\dfrac{k-9}{10k^2}$

(931) $\dfrac{-14m-16+6m^2}{(m-4)(m-3)}$

(932) $\dfrac{10r^3+3r-18}{2r^3(r-6)}$

(933) $\dfrac{9r^3+54r^2+2r+10}{6r(r+6)}$

(934) $\dfrac{4k^3-1+k}{2k^2(k-1)}$

(935) $\dfrac{b^2-b}{b-2}$

(936) $\dfrac{21r-19}{3(3r-4)}$

(937) $\dfrac{6r^2+28r+12}{(r+4)(r+3)}$

(938) $\dfrac{8m^2+51m-3}{6(m+6)}$

(939) $\dfrac{9a^2+66a+36}{(a+6)(3a+2)}$

(940) $\dfrac{11n-13}{5(3n+1)}$

(941) $\dfrac{2b^3-4b^2+5}{3b(b-2)}$

(942) $\dfrac{7b^2+5b}{(b-1)(b+1)}$

(943) $\dfrac{5r-15}{r(r-6)}$

www.math-knots.com

(944) $\dfrac{11v - 46}{6(v - 2)}$

(945) $\dfrac{3a - 10}{2(a - 3)}$

(946) $\dfrac{6k - 20}{(k - 4)(k - 3)}$

(947) $\dfrac{9x + 14}{(x - 2)(x + 6)}$

(948) $\dfrac{22r - 2}{5r(r - 1)}$

(949) $\dfrac{8 + 3r}{3r(r + 2)}$

(950) $\dfrac{42k + 18 + 6k^2}{(k + 6)(k + 3)}$

(951) $\dfrac{6x^2 - 69x - 50}{5(x - 5)(x + 2)}$

(952) $\dfrac{n + 24}{2(n + 6)}$

(953) $\dfrac{x - 19}{2(x + 5)}$

(954) $\dfrac{20a + 40}{3a(a - 10)}$

(955) $\dfrac{210b^3 - 180b^2 - 1}{3b(7b - 6)}$

(956) $\dfrac{7n + 12}{2(n + 2)}$

(957) $\dfrac{45x + 21}{8x(x - 7)}$

(958) $\dfrac{21}{(k - 5)(k + 2)}$

(959) $\dfrac{4p - 56}{(p - 10)(p - 6)}$

(960) $\dfrac{14r^2 + 25r + 24}{6r(r + 2)}$

(961) $\dfrac{-49x + 6 - 35x^2}{2x(7x + 10)}$

www.math-knots.com

(962) $\dfrac{-6x^3 - 6x^2 + 432x + 7}{(x - 8)(x + 9)}$

(963) $\dfrac{8a^2 - 56a - 3}{2a(a - 7)}$

(964) $\dfrac{-78n - 15 - 9n^2}{(n - 5)(n + 9)}$

(965) $\dfrac{140b^2 - 163b - 18}{12(7b - 8)}$

(966) $\dfrac{-2k - 2}{(k - 5)(k - 2)}$

(967) $\dfrac{7x^2 - 78x + 16}{7x(x - 2)}$

(968) $\dfrac{-x^2 + 2x + 4}{x - 2}$

(969) $\dfrac{78n - 8 - 5n^2}{10n(7n - 2)}$

(970) $\dfrac{9k + 15}{10(k - 3)}$

(971) $\{-2\}$

(972) $\{-1\}$

(973) $\{-1\}$

(974) $\left\{-\dfrac{5}{2}\right\}$

(975) $\left\{\dfrac{8}{5}\right\}$

(976) $\left\{\dfrac{24}{5}\right\}$

(977) $\left\{\dfrac{32}{5}\right\}$

(978) $\{5\}$

(979) $\{-1\}$

www.math-knots.com

(980) $\left\{\dfrac{23}{3}\right\}$

(981) $\{7\}$

(982) $\{2\}$

(983) $\left\{\dfrac{5}{4}\right\}$

(984) $\{-12\}$

(985) $\left\{-\dfrac{1}{4}\right\}$

(986) $\left\{\dfrac{16}{3}\right\}$

(987) $\left\{\dfrac{1}{5}\right\}$

(988) $\left\{-\dfrac{2}{3}\right\}$

(989) $\{-3\}$

(990) $\left\{-\dfrac{5}{4}\right\}$

(991) $\dfrac{5}{13}$

(992) $\dfrac{3}{5}$

(993) $\dfrac{3}{5}$

(994) $\dfrac{9}{41}$

(995) $\dfrac{3}{5}$

(996) $\dfrac{4}{5}$

(997) $\dfrac{3}{5}$

211   www.math-knots.com

(998) $\dfrac{4}{5}$

(999) $\dfrac{12}{13}$

(1000) $\dfrac{21}{29}$

(1001) $\dfrac{35}{37}$

(1002) $\dfrac{40}{41}$

(1003) $\dfrac{20}{29}$

(1004) $\dfrac{3}{5}$

(1005) $\dfrac{3}{5}$

(1006) $\dfrac{21}{29}$

(1007) $\dfrac{5}{13}$

(1008) $\dfrac{3}{5}$

(1009) $\dfrac{4}{5}$

(1010) $\dfrac{8}{17}$

(1011) $\dfrac{4}{3}$

(1012) $\dfrac{4}{3}$

(1013) $\dfrac{4}{3}$

(1014) $\dfrac{3}{4}$

(1015) $\dfrac{3}{4}$

212   www.math-knots.com

(1016) $\dfrac{40}{9}$

(1017) $\dfrac{21}{20}$

(1018) $\dfrac{3}{4}$

(1019) $\dfrac{3}{4}$

(1020) $\dfrac{4}{3}$

www.math-knots.com

www.ingramcontent.com/pod-product-compliance
Lightning Source LLC
Chambersburg PA
CBHW080840120626

46553CB00009B/2503